CW00506809

The Story of the Apple

The
Story
of the
Apple

Barrie E. Juniper

David J. Mabberley

TIMBER PRESS

Photographs by Barrie Juniper unless otherwise indicated.
Maps by Elanor McBay, The Drawing Office,
Department of Geography, University College London.
Frontispiece from *Giardino d'Agricoltura*, 1612, by
Marco Bussato, from Janson (1996).

Published in 2006 by
Timber Press, Inc.
The Haseltine Building
133 S.W. Second Avenue, Suite 450
Portland, Oregon 97204-3527, U.S.A.
www.timberpress.com
For contact information regarding editorial, marketing, sales, and
distribution in the United Kingdom, see www.timberpress.co.uk.

Printed in China

Library of Congress Cataloging-in-Publication Data

Juniper, B. E. (Barrie Edward)
    The story of the apple / Barrie E. Juniper, David J. Mabberley.
      p. cm.
    Includes bibliographical references and index.
    ISBN-13: 978-0-88192-784-9
    ISBN-10: 0-88192-784-8
  1. Apples. 2. Apples—History. I. Mabberley, D. J. II. Title.
    SB363.J86 2006
    634'.11—dc22
                                        2006011869

A catalog record for this book is also available from the British Library.

Dedicated to Cyril Dean Darlington
(1903–1981)
Sherardian Professor in the Botany School,
University of Oxford, who, through his book
The Evolution of Man and Society (1970),
taught us both to look at the interactions
between plants, animals, and
humankind

Contents

Color plates follow pages 80 and 144

Preface

THE APPLE has long been one of the most important fruits in the temperate regions of the world. It was apparent, certainly by the times of the Persians, Greeks, and Romans, that the apple is a remarkably convenient food source of great nutritional value. Human beings cannot make their own vitamin C, a deficiency rare among mammals. The apple is not only rich in this vitamin but, almost uniquely, can also be stored throughout a harsh winter or easily transported over long distances. Even completely dried, it loses little or nothing of its food value. In fact, there is an improvement in that bitter, almost completely inedible apples become palatable following quick drying and, in such a form, can be readily moved, handled, and stored yet remain virtually immune to cold, pests, and pathogens. There is evidence from Swiss lake dwellings that the almost inedible wild apple of Europe (*Malus sylvestris*) was treated in this way, as were the apples found on strings in Queen Pu-abi's tomb at Ur, about 4,500 years old.

Few if any crop plants resemble their wild ancestors. Compare the corncob of *Zea mays* with the small fruiting stalk of its wild ancestor, teosinte; consider the seedless banana, a triploid, having three sets of chromosomes, and thus sterile because it cannot produce viable gametes. Seedless fruits of all types, which have no genetic future without human intervention, abound, including most Smyrna-type figs, the 'Thompson Seedless' grape, and many derivatives of the mandarin (*Citrus reticulata*), for example, clementines and satsumas. Plants have also been selected by humans to exaggerate the development of roots or tubers (sugar beets, carrots, potatoes), leaves (lettuce), petioles (rhubarb), flowers (cauliflower), peduncles (the globe artichoke), to change their color (tomatoes and carrots), or to remove their poisons (broad beans)—the number of such distortions is legion. The huge fruiting stalks of the modern bread wheats derive from at least three different ancestors from at least two separate genera and have undergone two events of polyploidization, that is, multiplication of sets of chromosomes. In every case, people would scarcely recognize the wild parents of the crop plants.

The strawberry, too, is a polyploid hybrid, which arose in Europe, a cross between two very dissimilar wild species separated originally by the whole continent of North America whence they were both, at different times, imported. Doubling the chromosome number of the original sterile hybrid now allows it to reproduce sexually by seed. But the apple, with few exceptions—a small number of relatively modern sterile or semi-sterile triploids, including 'Ribston Pippin' (Plate 8), and a mere handful of commercially insignificant tetraploids—retains its original diploid chromosome complement of $2n = 34$. Indeed, the *Chromosome Atlas of Flowering Plants* (Darlington and Wylie 1955) listed just 18 triploids out of the whole gamut of many thousands of cultivars of the domesticated sweet apple (*Malus pumila*). The apple is still, for the most part, fully fertile, in sharp contrast to the now seriously threatened banana industry, which is widely represented by just two sterile cultivars: 'Gros Michel' and 'Cavendish'.

Since the time of Pliny the Elder (A.D. 23–79) and engaging the attention of such luminaries as Charles Darwin (1868), Alphonse de Candolle (1885), Nikolay Vavilov (1926, 1930, 1951, 1992), Elizabeth Schiemann (1932), and Peter Ucko and Geoffrey Dimbleby (1969), the origins of plants, particularly the major crop plants, have fascinated many plant scientists. But paradoxically—probably because of the relative inaccessibility of its homeland, both for geographical and political reasons—the origin of *Malus pumila,* unlike that of many other fruits and vegetables, long remained unknown.

For the greater part of what may be termed the scientific period of the study of plant origins, the cultivated apple was considered to be the result of hybridization involving the wild apples of the northern woodlands, that is, the European apple (*Malus sylvestris,* including *M. orientalis* of the Caucasus Mountains and Iran) and the Siberian crab (*M. baccata*). As recently as 1999 Li Yunong stated that the domesticated apple is a hybrid involving the wild apple formerly known as *M. sieversii,* and *M. sylvestris* (and *M. orientalis*). Earlier, John Loudon (1844) wrote with confidence, "Wild in woods and waysides in Europe. Cultivated in gardens, it is wholly, or conjointly with other species or races, the parent of innumerable varieties, termed generally, in England, cultivated apple trees; and in France, pommiers doux, or pommiers à couteau. We adopt the specific name *Malus* [that is, *Pyrus malus,* the name then used for the domesticated apple], to indicate what may be called the actual form, for the sake of convenience, though many of the cultivated varieties are derived not only from the wild apple, or crab, of Europe, but from the crabs of Siberia." In 1803 Moritz Borkhausen (see Korban and Skirvin 1984) believed that the progenitors of all *M. pumila* were *M. sylvestris* and the wild apples *M. dasyphylla* and *M. praecox.*

It would seem that they all were wrong.

Acknowledgments

B.E.J. was a Leverhulme Emeritus Research Fellow in the Department of Plant Sciences, University of Oxford, for the major part of the work for this book and wishes to thank all the following organizations and people for their contributions to the project:

The Head of the Department of Plant Sciences, Professor Christopher J. Leaver, C.B.E., F.R.S., who granted hospitality and all forms of services during the writing of this book; the Leverhulme Foundation for the funding of the major part of the work described here; and the Merlin Trust for travel grants.

His daughter, Sarah Juniper, who constantly encouraged him and offered all manner of suggestions and sources of information; Dr. Julian Robinson, now of the Dunn School of Pathology, Oxford, who carried out all the DNA analyses; Dr. John C. Smith of St. Catherine's College, Oxford, and Dr. Philip Durkin of *The Oxford English Dictionary* for the reinterpretation of so many strange words; Dr. J. N. Dimmick, Andrew Eburne, Professor Peter Franklin, Mr. Brendan McLaughlin, Dr. Gervase Rosser, Professor Gillian Sills, and Professor Michael Sullivan, all also of St. Catherine's College, for suggestions, for alerting him to, and for translating many early texts; Professor Robin Lane-Fox of New College, Thomas Braun of Merton College, Dr. Stephanie West of Hertford College, and Matthew Nicholls of St. John's College, Oxford, for translations and for drawing attention to many other Classical sources; Rabbi Eli Brackman, the Oxford Chabad, for advice on Hebraic texts; Stephen G. Haw of Chipping Norton, Oxfordshire, for the translation of Chinese texts; Lucinda Rumsey of Mansfield College for translations of Anglo-Saxon; Professor Rodney Thomson of the University of Tasmania for uncovering and translating the observations of William of Malmesbury; Professor John G. Dewey and Dr. Harold Reading, both of the Department of Earth Sciences, Oxford, for tutorials on Tian Shan geology; Professor Andrew Sherratt of the Ashmolean Museum, and Dr. Gail Preston of the Department of Plant Sciences, Oxford, both of whom told him about horses; Dr. Stephanie Dalley of Somerville College, Oxford, who told him about cuneiform writing; Dr. Malcolm Coe and Darren Mann of the Department of Zoology and the Oxford University Museum of Natural History, for education on dung beetles; Sandra Raphael of Oxford, who drew attention to many rare sources; Dr. Charles Jarvis and Professor W. T. Stearn, both of the Department of Botany in the Natural History Museum, London, for advice on nomenclature; Dr. Jörn Scharlemann of Oxford for several German translations and information on the von Richthofen family; Dr. Alistair Robb-Smith and Dr. P. Robb-Smith of

Woodstock for information concerning 'Blenheim Orange'; Dr. Oliver Rackham of Corpus Christi College, Cambridge, for information on the use of *apple* in English place-names; and Robert Franklin of All Soul's College, Oxford, who painstakingly read the whole work and suggested many improvements.

Dr. Ken Tobutt and the staff at Horticultural Research International, East Malling, Kent, for much help and for providing material of wild species from their collections; Frank Alston and Dr. Ray Watkins, both formerly of H.R.I. East Malling, for much help and advice; Dr. Emma-Jane Lamont, Alison Lean, and all the staff at Brogdale Horticultural Trust for assistance and encouragement of all kinds. Thanks, too, to John Lawson of East Hagbourne, Oxfordshire, for advice on, and making available, early texts on apple growing; Christopher Fairs of 'Bulmers' and Gerald Fayers of Gorleston, Great Yarmouth, Norfolk, for many sources of information and encouragement; the staff at the U.S. Department of Agriculture's Agricultural Research Service, Plant Genetic Resources Unit, Cornell University, in particular, Professor Philip Forsline, Professor Herbert S. Aldwinkle, Dr. Amy Szewc-McFadden, and Dr. Warren Lamboy for practical help and support in many ways; Professors Ralph and Lanna Lewin of San Diego for much advice; Dr. Lisa Karst of Rancho Santa Ana Botanic Garden, for help on diagrammatic interpretation; Dr. Anton C. Zeven of the University of Wageningen and Dr. Janneke Balk, now of the University of Marburg, for many Continental sources of information on the apple; John Rowe for much on the apple in Ireland; John Scott and Berrin Torolsan of Caique Publishing, with their particular expertise on Turkish matters, for much help on the apple in Turkey.

Professor Igor Belolipov and Dr. Luisa Samieva of the University of Tashkent for assistance in field collecting in Uzbekistan, Kazakhstan, and Kyrgyzstan; the staff of LES-IC, the Kyrgyz-Swiss Forestry Support Programme, in particular, Drs. Kaspar Schmidt and Sakir Sarimsakov and the Forestry Institute of the Republic of Kyrgyzstan, Jalal-Abad, for assistance in field collecting; Professor Nagima Ajtkhozina and Dr. Nazira Ajtkhozina of the Ministry of Sciences and the University of Almaty, Kazakhstan, for assistance in field collecting; Professors Lin Pei Jun, Li Jiang, and Tan Dunyan, and Drs. Pi Er Dong, Gao Jien, Liao Kang, Zhao Jinchun, and Cong Peihua, and particularly, Dr. Zhi Qin Zhou, for assistance in field collecting in China; Dr. V. V. Ponomarenko of the Vavilov Institute, St. Petersburg, and Professor Valentina Vereshchagina of Perm State University of Russia for advice on apple taxonomy and the distribution of wild apples in Russia.

Andrew Liddell and John Baker of the Department of Plant Sciences, and Jamie Keats of St. Catherine's College, Oxford, for assistance in preparing the scanned images, and Elanor McBay of University College London, who drew all the maps. In particular, thanks to Rosemary Wise of the Department of Plant Sciences for most of the artwork. Above all, thanks to Dr. Stephen Harris, Druce Curator of the Oxford University Herbaria, Department of Plant Sciences, for his constant support, advice, enlightened discussion, field companionship, enthusiasm, and immense tolerance of idiotic questioning over many years.

D.J.M., who not only studied at the feet of B.E.J. when an undergraduate at St. Catherine's College, Oxford, but also helped move B.E.J.'s burgeoning apple tree collection in the late 1960s, wishes to thank B.E.J. for the invitation to join in the writing of this book. He is also grateful to Steven McKay of the University of Minnesota for providing pamphlets and information on apples in North America and to Claudio Pericin for information on apples in Istria, Croatia. Both B.E.J. and D.J.M. thank Timber Press for encouragement throughout the gestation of the book.

The story of the apple ranges over a large area. The republics, regions, cities, districts, rivers, and mountains of the Middle East and Inner and Central Asia have, over millennia, suffered many name changes as successive waves of invaders imposed their rule. The names used in the text and on the maps here are neither to be considered definitive nor based on some consistent authority—they are simply those in common use. Some names have changed considerably. For example, what was once Vyernyi has become successively Verney, A-li-ma, Almaly, Alma-Ata, and is now Almaty, Frunze now Bishkek, and Constantinople now Istanbul.

The apple story is intricate, and its investigation has led to research in a wide range of disciplines with their own technical languages. To help readers, synopses are provided at the beginnings of the main chapters, and the whole story is summarized in the dénouement.

CHAPTER 1

What are apples?

THE DOMESTIC OR SWEET APPLE is correctly called *Malus pumila*, the name of a species in the family Rosaceae, tribe Pyreae, comprising some 11 genera and 620 species. The often repeated theory of the origin of the tribe through hybridity between species in two other groups is replaced by the idea that the Pyreae arose within the *Spiraea* group in North America and spread westward to China, where species survived the glaciations until later recolonizing North America.

Hybridization, also mooted as significant in the evolution of the domestic apple itself, seems rare between *Malus pumila* and wild apple species and not to have been a factor. Remarkably, *M. pumila* has almost no endosperm, a feature that may help explain this. By comparison with the bird-dispersed fruits of other apple species, its fruit is very large, suggesting that its original dispersal agents were large herbivorous mammals.

The genus *Malus* probably arose in the Tertiary in southern China and spread through a continuous corridor of temperate forest as far as western Europe, a passageway now greatly fragmented, one relict being the fruit forest of the Tian Shan of Inner and Central Asia. It is argued that the unique combination of physical and biological characteristics found in that range was the driving force for the evolution of *M. pumila*.

TODAY, apples are considered to belong to the genus *Malus*. All plant species are arranged according to a taxonomic hierarchy in which they are gathered into larger groups or subdivided into smaller ones to account for variation. Each of these groups is given a particular name. For example, the taxonomic hierarchy for the apple 'Bramley's Seedling' is, in full, family Rosaceae, subfamily Spiraeoideae, tribe Pyreae, genus *Malus*, section *Malus*, series *Malus*, species *M. pumila*

Miller (Philip Miller having first described the species according to the rules of botanical nomenclature in 1768), cultivar 'Bramley's Seedling'. Thus the rose family, Rosaceae, includes the genus *Malus*. *Malus* species are grouped into sections and series, and species are sometimes divided into infraspecific groups: naturally occurring subspecies and varieties in addition to "cultivated varieties," known correctly as cultivars. A current classification of *Malus* species is given in the Appendix (adapted from Phipps et al. 1990).

Note that the words for genera and species are shown in italics (for example, *Malus baccata*) whereas known or suspected hybrids are printed with a multiplication sign × added, for example, *M.* ×*floribunda*, thought to be a hybrid between *M. sieboldii* (included in *M. toringo*) and *M. baccata* (Plate 5). Note also that the correct scientific name of the cultivated apple, testifying to the very long association with humans, has long been debated. There were early arguments for *M. communis* (see Lamarck and Poiret 1804); names used relatively recently include *Pyrus malus, P. malus* var. *paradisiaca, M. pumila, M. sylvestris, M. sylvestris* var. *mitis,* and *M. domestica,* the last sometimes treated as a hybrid rather than a species. Although the last name has been brought into current use (Korban and Skirvin 1984), it is illegitimate according to the rules of botanical nomenclature, and *M. pumila* (Plates 1 and 2), as long used by various authors concerned with the evolution of the crop plant (for example, Zohary and Hopf 2000), has been shown to be correct (Mabberley et al. 2001).

Apple relatives

Fossil plants that appear to be the antecedents of what has become the large modern-day family Rosaceae seem to have emerged as recognizable entities sometime between 50 million and 40 million years ago (Table 1). Today, there are 17 more or less clearly distinguishable tribes of Rosaceae (Kalkman 2004), including, for example, Roseae, comprising the genus *Rosa* (the roses); Rubeae, with *Rubus* (blackberries and brambles); Potentilleae, with *Potentilla* (including *Fragaria,* strawberries; Mabberley 2002); Pruneae (almonds, cherries, plums, and apricots, the so-called stone fruits); Spiraeeae, with *Spiraea;* Crataegeae, with *Crataegus* (hawthorns); and relevant to this book, Pyreae (formerly called Maleae or Pomoideae and including apples, pears, and quinces—the so-called pome fruits; Huckins 1972, Phipps et al. 1991, Aldasoro et al. 2005).

Distinguishing them from the other tribes, the Pyreae have radially symmetric (actinomorphic) bisexual flowers with the floral parts in fives. The flowers characteristically have two to five carpels (the parts of the ovary that contain the

TABLE 1. Geological time and the apple

PERIOD	EPOCH	MILLION YEARS AGO	EVENTS
Quaternary	Holocene	0.008–0	postglacial to present
	Pleistocene	1.8–0.008	Neolithic transition
Tertiary	Pliocene	5.3–1.8	glaciation and desertification and isolation of the Tian Shan
	Miocene	23.8–5.3	emergence of the Tian Shan
			members of the Pyreae recorded in fossils in North America
	Oligocene	33.7–23.8	
	Eocene	55.5–33.7	possible origin of the tribe Pyreae, the apples and their allies
			Indian landmass collides with the northern landmass
	Paleocene	65–55.5	Rosaceae are recognizable as such in the fossil record
Cretaceous		145–65	angiosperms diverge from other seed plants about 120 million years ago

ovules) enclosed in a swollen flower stalk (receptacle) and topped by the rest of the flower, resulting in a so-called inferior ovary (Phipps et al. 1991). This ovary and surrounding tissues mature into the characteristic pome, the most striking characteristic of the tribe and defined as a fleshy, indehiscent, so-called false fruit formed from a flower of which the true fruit is surrounded at maturity by an enlarged floral tube or a fleshy receptacle, or both (Rohrer et al. 1994).

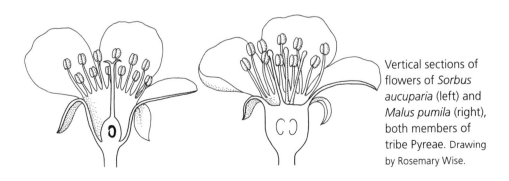

Vertical sections of flowers of *Sorbus aucuparia* (left) and *Malus pumila* (right), both members of tribe Pyreae. Drawing by Rosemary Wise.

The Pyreae comprise 11 genera (Kalkman 2004), though as many as 24 have been recognized (Aldasoro et al. 2005), and probably about 620 species. Precisely where or when this group diverged from other Rosaceae it is as yet impossible to say, but apparently somewhere in the northern hemisphere and, according to the current orthodoxy, on the central-eastern Asian landmass in the Tertiary (Li Yunong 1989, 1996a, b; see Aldasoro et al. 2005 for a comprehensive review of the fossil evidence). A late Eocene origin for the group has been suggested (Table 1), but it is admitted that this is conjectural (Phipps et al. 1991). Whatever the nature of this particular event, it would appear to have manifested itself in the evolution of numerous prominent garden and crop genera of Rosaceae, namely, *Amelanchier* (serviceberries), *Aronia* (chokeberries, and including *Photinia* and *Stransvaesia*), *Crataegus* (hawthorns), *Eriobotrya, Malus, Pyrus* (pears), *Rhaphiolepis*, and *Sorbus* (mountain ashes).

The Roseae have a number of characteristics and tendencies not found at all, or only relatively uncommonly, in the Pyreae—for example, a tendency toward the climbing habit, while all have numerous carpels, each with only one ovule.

The Pruneae more nearly resemble the Pyreae in habit, leaf shape, inflorescence, and characteristics of sepals and petals than they do any other group of Rosaceae. Moreover, unlike any other Rosaceae, plants in these two tribes synthesize amygdalin (vitamin B_{17}), a cyanogenic glycoside. Hydrolysis releases the cyanide, which can be detected in crushed almonds or apple pips (seeds) but is not normally present in quantities dangerous to human beings. It may, though, be enough to deter a seed-eating bird. However, there are certain morphological and anatomical characteristics in which the Pyreae resemble the Spiraeeae more than they resemble the Pruneae, such as the presence of two to five carpels, sometimes with several or numerous ovules (Kalkman 2004; a comprehensive review of the earlier literature can be found in Huckins 1972).

It has been argued that many genera, subfamilies, and even families of flowering plants have arisen through polyploidy through the combining of unreduced (diploid) gametes or, more rarely, via doubling of sets of chromosomes following hybridization, rendering an initially sterile hybrid capable of producing viable gametes and thus producing viable seeds. The phenomenon is common in the widespread families Rosaceae, Rubiaceae, Compositae, and Gramineae (Darlington and Moffett 1930, Stebbins 1950, Darlington 1956, Grant 1971, Phipps et al. 1991, Evans and Campbell 2002). The Pyreae and their allies, with almost no exceptions, have a base chromosome number $x = 17$. This not being an even number suggests that if polyploidization occurred in the past, chromosome number modification (loss or addition of chromosomes, possibly by fragmentation) has subsequently taken place.

There are several hypotheses, dating back to the 1930s, that seek to explain the evolution of the Pyreae. These proposals can be divided into two groups, those that assume an origin from a single existing group versus those that base the evolution on hybridization between two distinct groups. Darlington and Moffett (1930) put forward a proposition to explain the apparent ancient polyploid origin of Pyreae (what they called Pomoideae). They found that in the species with 17 bivalents (17 pairs of chromosomes), the complement of chromosomes could be divided into three groups of three bivalents each and four groups of two each. They proposed that Pyreae originated from an original diploid species with base chromosome number $x = 7$ by doubling of the entire set plus the addition of a third partial set comprising three of the original seven bivalents, thus $x = 7 + 7 + 3 = 17$. Contrary to this model, initially proposed by Darlington for *Pyrus* (which at the time was considered to include *Malus*), Sax (1931) suggested that the Pyreae and their allies are actually allopolyploids, polyploids derived from two distinct ancestors evolved from primitive Pruneae.

In the light of modern knowledge, more than one interpretation of these two proposals can be considered. Since the basic chromosome number most commonly found in the Spiraeeae is $x = 9$, and that of the Pruneae is $x = 8$, very tentatively and in the absence of a modern sophisticated karyotype analysis the following evolutionary event can be proposed: diploid Pruneae ($x = 8$) × diploid Spiraeeae ($x = 9$) = allopolyploid Pyreae ($x = 17$).

A second, more recent hypothesis based on modern molecular techniques (Morgan et al. 1994, Campbell et al. 1995), along with nonmolecular data, focuses on four rare, ancient, but extant genera of Rosaceae: *Kageneckia*, *Lindleya*, *Porteranthus*, and *Vauquelinia*. The current distributions of all four now lie in the southern part of North America, Central America, or South America. Their exact taxonomic position in the family has long been a puzzle. The new evidence seems to place them all near a large group that includes Pyreae rather than in a heterogeneous assemblage that includes Spiraeeae, the whole analysis supporting a strong evolutionary link between modern Pyreae and Spiraeeae. Indeed, the early development of the gynoecium and ovule in *Vauquelinia* is rather similar to that in Pyreae. *Kageneckia* and *Lindleya* have a base chromosome number $x = 17$ (like that of Pyreae), *Vauquelinia* has $x = 15$, and *Porteranthus* $x = 9$. If *Porteranthus* can be confirmed as allied to the Pyreae, then along with the evidence from *Kageneckia* and *Lindleya*, linking the Pyreae more firmly with the Spiraeeae, an autopolyploid origin of the Pyreae from an early spiraeoid group becomes more plausible (Morgan et al. 1994): early diploid Spiraeeae or allies ($x = 9$) undergoing autopolyploidy ($x = 18$), *auto-* signifying one rather than two ancestors, followed by reduction in chromosome number to $x = 17$.

Evans and Campbell (2002) sequenced the DNA of a gene, called GBSS1 or *waxy*, in *Malus* that is responsible for the synthesis of starch. This gene is present as at least two copies in all Rosaceae, and as four in the Pyreae. *Gillenia*, sometimes construed as including the genus *Porteranthus*, is native to the southeastern United States and has a base chromosome number $x = 9$. Evans and Campbell hypothesized that somewhere in North America, ancestors of, or plants related to, present-day *Gillenia* gave rise to the Pyreae. Like Kubitzki (in Kalkman 2004), they rejected the hybridization theories that have occupied the scientific literature for so long. Indeed, more recent DNA research places Pyreae firmly in subfamily Spiraeoideae along with six other tribes, the rest of the family divided into subfamilies Rosoideae and Dryadoideae (Eriksson et al. 2005).

If Evans and Campbell (2002) are correct, it has to be assumed that long ago the early ancestors of the Pyreae migrated west from North America, presumably across the Bering land bridge (now submerged under the Bering Strait; Map 1), to a preglacial or interglacial refuge in what is now central and southern China. Whatever the ultimate origin, the center of distribution and most recent evolution, at least at the species level, appears to have remained within central-eastern Asia. Phipps et al. (1991) argued that in the Eocene the major genera of the Pyreae and their allies very rapidly took on the appearances we would recognize today. Did then representatives of genera such as *Amelanchier, Crataegus, Malus,* and *Sorbus,* but oddly enough not *Pyrus,* migrate east, perhaps more than once, and probably again across Beringia and perhaps through a temperate forest corridor extending around the northern hemisphere? These pioneers must have penetrated into what is now North America and indeed can be identified as distinct genera there by the Miocene if not by the Oligocene (Table 1; Boufford and Spongberg 1983). The dispersal of plant genera or species from one landmass to

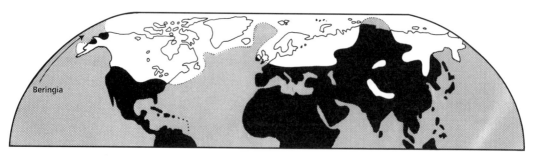

MAP 1. The extent of the ice cover in the most recent glaciations is shown in white. Continental and lake boundaries are, with this ice accumulation, almost completely meaningless but are reproduced for guidance. An ice-free corridor, Beringia, now the Bering Strait, from time to time joined Asia and North America.

another is a highly controversial subject, however, and a revision of these speculations in the future is likely.

Amelanchier, Aronia, Crataegus, and *Sorbus* appear to have flourished in North America and, in numbers of species, have proliferated, often much more spectacularly than their relatives in Inner, Central, and eastern Asia. *Mespilus,* on the other hand, has only two species, the medlar (*Mespilus germanica*) of Europe and the much more recently discovered *M. canescens,* known only from a small population in Arkansas (Phipps 2003), though considered by some to be merely a hybrid between the introduced medlar and a hawthorn. Of the Pyreae and Crataegeae, only species of *Cotoneaster* (to Africa), *Hesperomeles* (to South America), and *Aronia* (including *Photinia,* to Java), all Pyreae, have penetrated, to a limited extent, south of the equator.

Under experimental conditions, intergeneric hybrids are possible between virtually all the genera of Pyreae and their allies. ×*Amelosorbus,* ×*Crataemespilus,* ×*Pyromalus,* ×*Pyronia,* ×*Sorbaronia,* and ×*Sorbopyrus* (Rehder 1926, Sax 1931) have been recorded, though the frequency of such hybridization casts some doubt on the validity of the distinctness of the genera (Walters 1961, Mabberley 2002). None of these hybrids, except for the hybrid between a hawthorn and a medlar, ×*Crataemespilus grandiflora,* appears to be of any economic value whatsoever. That plant is a useful ornamental large shrub or small tree grown in gardens in the temperate zone.

In addition, there are very strange, stable, graft hybrids (not true hybrids since there is fusion of tissues rather than gametic fusion, so indicated by + rather than ×): +*Crataegomespilus dardarii,* sometimes known as the Bronvaux medlars, involving the hawthorn *Crataegus monogyna* and the medlar *Mespilus germanica.* The outer layers of tissue are those of the medlar, though the fruit shape is that of the hawthorn. The gametes (eggs and pollen) are derived solely from the medlar. The graft hybrid has arisen more than once, so there are a number of cultivars, which can be propagated by cuttings.

There is no obvious barrier, except as limited by floral timing, to hybridization either between species of different genera within the Pyreae or between species of *Malus.* Natural distributions comfortably overlap. Species hybrids, both spontaneous and man-made, are common in many other commercial crops and ornamentals, such as *Rosa* (roses), *Triticum* (wheats), *Trifolium* (clovers), *Potentilla* (including *Fragaria,* strawberries), and probably *Pyrus* (pears), but in spite of frequent assertions to the contrary, interspecific hybridization appears to play little or no part at all in the polymorphism of the sweet apple.

In immediate prehistory or just within the period of recorded history, small populations of *Malus pumila* were carried westward over thousands of kilometers.

During that time there has been more than adequate opportunity for hybridization with, for example, the isolated species of the Middle East and Europe, among which are *M. trilobata* of the eastern Mediterranean, *M. florentina* of Italy, Greece, and Turkey, and *M. sylvestris* (Plate 3) of Europe, the latter including *M. orientalis* of the Caucasus Mountains and Iran. But it has to be concluded that these species, probably at different times, came to an evolutionary halt on isolated branchlines and are now unable to achieve fecund hybridization with any other species. In Kyrgyzstan, *M. pumila* overlaps with the local but distinct *M. kirghisorum* (Plate 4 and Map 5), but although hybrids are reported, they seem to be rare and not to have spread.

The phrase "natural hybrids occur," although not wholly lacking, seems to be singularly absent in descriptions of apple species, particularly so in *Malus pumila*. Only a few modern cultivars of ornamental crab apples (Chapter 7) appear to belie this statement. The Soulard crab, *M. ×soulardii* (Anderson 1952), common near habitation in Missouri, is believed to be a cross between an imported sweet apple and the local native, *M. ioensis. Malus ×dawsoniana* is thought to be a hybrid of a sweet apple and the Oregon crab, *M. fusca. Malus ×floribunda* of Japan is considered to be a cross between *M. sieboldii* (included in *M. toringo*) and *M. baccata*. But these assertions remain to be checked by modern methods. However, it is widely claimed that hybridizations have taken place, particularly in the evolution of the sweet apple (Borkhausen 1803 to Li Yunong 1999). Although such crosses have been attempted for agricultural purposes, no confirmed example has led to significant advance, contrary to crosses in so many other genera and species (Dennis 2003).

Conversations with plant breeders familiar with, or directly involved in, plant breeding between the World Wars and shortly after World War II suggest that hybrids between *Malus pumila* and *M. sylvestris* (the latter including *M. orientalis*) were attempted but came to nothing. DNA work in Belgium confirms, through the use of arbitrary fragment length polymorphisms and microsatellite markers (Chapter 2), that notwithstanding ample opportunity, hybrids between the wild *M. sylvestris* and the cultivated sweet apple are very rare; moreover, hybrids between *M. sylvestris* and exotic ornamental *Malus* species were not detected at all (Coart et al. 2003).

Although polyploidy in both its auto- and allo- forms is very common in many crop plants, it does not appear to have played a role in the recent evolutionary history of the domesticated sweet apple. However, it must be pointed out that there is a small number of apple cultivars, mainly "cookers" that are sterile or semi-sterile triploids ($3x$), presumably all of which arose by fusion of an unreduced gamete ($1x$) with a normally reduced germ cell ($2x$) during meiosis (the formation of gametes). These include 'Baldwin', 'Belle de Boskoop', 'Blenheim Orange', 'Bramley's Seedling', 'Gravenstein', 'Jonagold', 'Reinette du Canada', 'Rhode Island Greening', 'Ribston Pippin' (Plate 8), 'Warner's King', and 'Washington'. Most, probably starting with 'Baldwin' (1740), 'Bramley's Seedling' (about 1809), and 'Belle de Boskoop' (1856), are of comparatively modern origin. The oldest appears to be 'Ribston Pippin', possibly raised before 1688 but more probably from the very early 18th century. There are said to be tetraploid ($4x$) mutants of the American 'Rhode Island Greening' and 'Perrine Giant Transparent' (Morgan and Richards 1993; the latter also known as 'Giant Transparent', 'Giant Yellow Transparent', and 'Grandparent') in addition to the colchicine-induced tetraploid developed in Ontario, Canada, 'Hunter Sandow', but these play little part in the commercial market, and in any case their alleged ploidy level should be rechecked using modern methods.

Apple trees are hermaphrodites, carrying both sexes within the same flower. They are usually self-incompatible (unable to fertilize themselves), with self-pollen tube growth stopped in the style (Janick et al. 1996). Almost all the cultivars of the sweet apple are self-incompatible to a greater or lesser extent; some are completely so, though many will set a small amount of fruit by self-pollination. Through unconscious selection, some of the more modern cultivars will set some fruit, but current commercial apple production is inconceivable without recourse to a cross-pollination strategy. In the British Isles only the cultivars 'Allington Pippin', 'Lord Grosvenor', and 'Stirling Castle' will give any significant crop as a result of self-pollination. In general commercial practice, the identity of the pollen parent, provided that it is a species of *Malus,* and the precise timing of that pollination event are of no significance. Nonetheless, there is evidence of some cross-incompatibility between 'Cox's Orange Pippin' and its seedlings, for example, 'Saint Everard', 'Ellison's Orange', and 'Laxton's Superb' (Hall and Crane 1933). If the identity of the pollen parent is relatively unimportant in *M. pumila,* this is not wholly true of plants in general (Marshall and Oliveras 2001). But in the high-density fruit forest of Inner and Central Asia, where *M. pumila* may sometimes make up 80% of the trees (Plate 17), these subtle differences of cross-pollination are probably of little significance.

Incompatibility alleles (alleles are different versions of a gene) form proteins that prevent pollen of a particular genotype, or genetic constitution, from germinating on the stigmatic surface of a neighboring plant that does not possess a complementary set of these alleles. In Pyreae, incompatibility is controlled by a series of genetic sequences known as S (self-incompatibility) alleles. If the apple possesses an S-1 allele it will neither successfully fertilize nor be fertilized by another S-1 carrier. Thus the more S alleles possessed, the wider a donor or recipient must seek for a partner. The first work on the control of compatibility in *Malus* and the identification of 11 alleles was by Kobel et al. (1939), and there are as many as 25 such alleles (Bošković and Tobutt 1999).

Over the vast periods of time available to species in the Tian Shan (also spelled Tien Shan), roughly 1,000 times greater than that enjoyed by species re-invading the British Isles or parts of North America after the last glaciation, for example, one might speculate that incompatibility alleles could have accumulated. Perhaps this accumulation might have come about by very early hybridizations. Being advantageous in promoting outbreeding, such alleles would not easily be selected against; they would have promoted wider and wider outbreeding where trees of the same species in close proximity was the norm. If incompatibility had developed to a high degree, this might explain why *Malus pumila*, seemingly unique among *Malus* species in particular and many other crop genera in general, has shown such diversity on the one hand and such a marked reluctance to hybridize on the other. The astonishing degree of diversity or heterozygosity resulting from a single *M. pumila* × *M. pumila* cultivar cross has been illustrated (Brown and Maloney 2003: pl. 3.2). Almost every size, color, and shape seen in commercial apples, as well as variety in the habit of mature trees, is released from one single parental event. A similar variety is found in the natural fruit forest of the Tian Shan.

Apple fruits, flowers, and seeds

As with other Pyreae flowers, those of *Malus* are not specialized for pollination by any one group of insects. The simple, usually gently fragrant (to humans), and readily accessible flowers attract a range of pollinators seeking a reward of nectar or pollen, or both. The whole of the Tian Shan must have been a paradise for the many species of wild bees. They would have fed on a wide range of fruiting trees, predominantly but not entirely members of the family Rosaceae, and meadow herbs in due season, giving rise to what is now an important honey industry. In places, their aerial concentration is so thick one seems almost to breathe bees. But the industrial worker is *Apis mellifera*, the honeybee, which may indeed have originated in the area but thanks to its adaptation to human requirements is probably now the most widespread of all bee species (Plate 19). There is evidence that other species, however, including the solitary bees sometimes known as mason bees, similar to the northwestern European natives *Osmia rufa* and *O. cornuta*, may have played a much greater role in the pollination of apples in the past and thus in the evolutionary process.

Apis mellifera is not well adapted to pollination of the apple flower. Frequently, because of the impeding stamens, honeybees bite through the base of the calyx to reach the nectar, thus negating the cross-pollination process. In contrast, mason bees, seeking only the pollen—they store only pollen and not nectar—come in direct contact with the sexual parts of the flower. Honeybees and bumblebees have their pollen baskets (scopae), which are concave surfaces with stiff bristles, on the outer faces of the hind legs. Leaf-cutter bees and mason bees, including species of *Osmia*, have a scopa in the form of stiff hairs on the underside of the abdomen, so that they are much better positioned for the transfer of pollen in an apple flower. Moreover, *O. rufa*, the red mason bee, is active in the spring before the overwintering honeybee colonies have built up to full effectiveness. Thus it is estimated that, given all these advantages, one red mason bee can do the work of 120 honeybee workers (O'Toole 2000). But the rare ability of strains of *A. mellifera* to accumulate that very rich foodstuff, honey, has resulted in their swamping the pollination scene virtually throughout the world.

Of the apple fruit, the pome, the piece eaten is essentially a womb, a fleshy envelope developed from tissue at the base of the flower and wrapped around the flimsy, inedible true fruit. There are from three to usually five carpels, each containing two seeds, or one by abortion. The seeds usually have a thin lining of endosperm, a tissue unique to flowering plants—it has three parents, one male nucleus from the pollen grain and two from the egg sac, but never leaves any descendants. *Malus pumila* has almost no endosperm, distinguishing it from other apple species. It is now believed that the endosperm may serve several other key functions in addition to nourishing the embryo, including that of a fertilization sensor, detecting and aborting the fertilization products of incompatible crosses or crosses between unrelated species (Costa et al. 2004). This absence of endosperm may be a factor in the inability of *M. pumila* to form hybrids with other *Malus* species, but this whole area of genetic imprinting and epigenetics is still highly speculative and barely explored.

The fruit will not normally develop beyond a pygmy stage until more or less complete pollination has been achieved. Such pygmy fruits are usually discarded through premature dehiscence—the so-called June drop in an English orchard. Partial pollination generally results in misshapen fruits. Parthenocarpy, the development of a fruit without fertilization of the egg cell, was noticed in apples as long ago as 1685

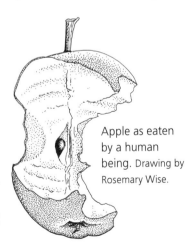

Apple as eaten by a human being. Drawing by Rosemary Wise.

("A Lover of Planting" 1685, Juniper and Juniper 2003). A small number of seedless apple cultivars, such as 'Wellington Bloomless', 'Spencer Seedless', and 'Rae Ime', have been selected, but on the whole these are of little or no value as commercial fruit crops. Nevertheless, the financial goal of a high-quality, seedless apple along the lines of the common and successful cultivars of banana, tangerine, or table grape has stimulated research into the genetic basis of the phenomenon.

One of the proteins, PISTILLATA, manufactured in the second and third whorls of simple angiosperm flowers, converts organs that would otherwise become carpels and styles into petals and anthers, respectively. This protein is encoded (stored) and at the correct moment expressed in the well-studied mouse-ear cress, *Arabidopsis thaliana*, in the cabbage family (Cruciferae), by the gene *PISTILLATA*. A region resembling this gene has been found in the genome of 'Granny Smith' apples. An extra length of DNA was discovered in the gene region of the 'Rae Ime' apple that renders *PISTILLATA* inoperative (Yao et al. 2001). Similar defects were later found in the *PISTILLATA* region of 'Wellington Bloomless' and 'Spencer Seedless'. Therefore, sometime in the future, orchards of seedless apples and redundant bee colonies could become the norm.

The seeds of Pyreae will not germinate without two requirements being met. First, there must be separation of the seeds from the basic tissue, particularly the placental tissue of the apple core, which contains germination inhibitors; apple pips and many other fruit seeds will not germinate if left in the core (Evenari 1949, Herb Aldwinckle pers. comm. 2001). Second, the seeds usually need a substantial cold-chill, a long period in the free-seed stage at or near the freezing point of water, after which, but not without which, germination is possible. This cold-chill requirement is a very common feature of seeds of woody plants and is not confined to apples. It obviously has great survival value where a brief period of warmer weather in a severe winter may imperil prematurely germinating seedlings.

In some plants, dormancy is found in only part of the embryo. For example, acorns, the fruits of oaks, *Quercus* spp., of the oak family (Fagaceae), will produce a short, immature, embryonic root, the radicle, if they are planted directly after shedding, but the epicotyl—the embryonic aboveground growth—does not grow until the cold-chill requirement is met. Other plants, such as lily-of-the-valley, *Convallaria majalis* (Asparagaceae), require one period of chilling for the radicle to emerge and a second one to stimulate the growth of the epicotyl. Hence, in those species, seedlings will not appear above the ground until the second season after planting.

Such sustained periods of cold, 60 continuous days or more at 2°C (36°F) or thereabouts, are consistent features of the climate in Inner and Central Asia. They also prevail in the other strongholds of *Malus* species: central China, the

Caucasus Mountains, and continental North America (for introduced apples). Some populations of *M. pumila* may require as long as 200 days of chilling for effective germination (Forsline et al. 2003). In consequence, the mild winters of contemporary Britain, for example, stimulate very little seed germination, though the practice of shipping, transferring, and storing fruits and then displaying them in refrigerated cabinets in petrol service stations does seem to be leading to local germination of feral apple trees along motorways.

Extended cold periods are also rarely experienced in the milder climates of the maritime regions of western continental Europe, coastal North America, Australia, or much of South Africa, where there has been more recent selective breeding for low-chill cultivars (those with a shorter requirement for winter chilling). Thus on the one hand, apple breeders in South Africa are seeking shorter cold-chill requirements (Forsline et al. 2003), whereas on the other, longer chilling requirements are sought to provide later-blooming cultivars with the capacity to avoid spring frosts in more continental parts of the world such as Russia. As a result of the erratic distribution of the cold-chill requirement throughout the world, comprehensive textual guidance is available for would-be breeders (Crossley 1974, Westwood 1995, J. W. Palmer et al. 2003, Wertheim and Webster 2003). Stratification—the placing of seeds in layers of moist sand, peat, or something similar for extended periods at low temperatures—enhances germination. There are several advantages to the use of seeds, for example, in rootstock production, not the least of which is that almost no viral, fungal, or bacterial pathogens are transmitted via seeds.

Although not as well documented or researched as the cold-chill requirement or stratification, another way to enhance germination is through scarification. This is the scratching or thinning of the seed coat, the testa, by physical or chemical means, or both. This may take place in the jaw, crop, stomach, or intestine of a wide range of animals. Almost all hard-coated seeds benefit from such treatment by scrubbing away any residue of germination inhibitors and rendering the surface covering weaker and more permeable to water and oxygen. This effect is considered later in the discussion of the role of various animals in apple dispersal.

For successful bud formation, breaking of dormancy, and emergence, apple trees must also be subjected to extended cool periods. Optimal chilling efficacy would appear to be 7.2°C (45°F), with zero response below −1.1°C (30°F) or above 15°C (59°F). These criteria seem to be the limiting factors in determining the effective fruiting distribution of *Malus pumila*, roughly between 25° and 52° north (to more than 53° north at Edmonton, Alberta, Canada; Steven McKay pers. comm. 2005) in the northern hemisphere, but there are many cultivar deviations and climatic anomalies modifying this general rule, and the subject is exceedingly complex. The presence or absence of cold-chill areas in the discussion of late-stage colonization by the apples of southern temperate zones is considered in Chapter 6.

Apple dispersal

From the history of glaciation and desiccation, it would seem that the evolving apple lineages followed mountain chains and the banks of rivers and streams that originate from them, to the plains. Thus in the Old World, apple populations can be found in, among other places, the Altai Shan of western Mongolia to Kazakhstan, the Tian Shan, the Himalaya, the Caucasus Mountains, the Taurus Mountains of southern Turkey, the Carpathians of eastern Europe, the Alps, and the Pyrenees (Maps 3, 4, 8, and 9).

The wild European apple, *Malus sylvestris* (Plate 3), is widely distributed in eastern, central, and western Europe but seems nowhere to be common (Remmy and Gruber 1993). In Belgium, for example, it is designated as endangered (Coart et al. 2003). In the British Isles, where it is as rare, it reaches to 59° north in the Shetland Islands, as it does in Sweden, while in Finland it is found as far as 62° north. In Russia it extends to 60° north, but farther east, in Siberia and northern China, it is replaced by the Siberian crab, *M. baccata* (Plate 5). It looks as though 60–62° north is the extreme limit for any wild species of *Malus*. The sweet cultivars of *M. pumila*, on the other hand, seem to be limited roughly to a band between 25° and 55° north in the northern hemisphere.

There is a great deal of confusion between the genuine wild European apple, *Malus sylvestris*, and feral seedlings of *M. pumila* imports. The only readily observable floral distinction is that at the flowering stage, the calyx lobes of *M. pumila* are distinctly hairy on both surfaces, whereas in *M. sylvestris* only the inner one is hairy, and the outer smooth (Plate 3). Nevertheless, these two species can be found growing near one another and are often confused. Although we now believe that *M. sylvestris* played little if any role at all in the evolution of the domestic apple, it is nevertheless a widespread and familiar species. In the absence of detailed work on wild *M. pumila*, some interesting characteristics of the well-studied *M. sylvestris* may serve as a paradigm for an investigation into apple species in general.

In parts of pre-agricultural Europe, *Malus sylvestris* was an important if infrequent small-tree element in open woodland and semi-natural grasslands. Although awareness of this has increased, little appears to be known about the dispersal of the species in nature. Field study suggests that large domestic herbivores may be major dispersers of seeds as well as providers of suitable seedbeds for germination, but which herbivores? Cattle, in particular the original wild bison (*Bison bonasus*), and possibly some of the larger deer would appear through trampling by their sharp hooves to contribute to germination success by driving the tough-coated apple seeds through a compact turf layer, similar to the role of deer in the establishment of *Fritillaria meleagris* (Liliaceae) in flood meadows in northwestern Europe. Wild boar (*Sus scrofa*) were present throughout all western Europe until relatively recently, having been exterminated from England in the 14th century and Scotland by the early 17th. They probably roamed through the transcontinental stretch of the Tertiary forest corridor and could have promoted dispersal in their digging, with tusks, snout, and hooves, in search of roots and tubers, and laying waste, as if with a plow, to large areas of turf. They must have served, indirectly, to spread and bury seeds of all kinds, as other pig species still do in tropical forests today (Mabberley 1992). The emergence and subsequent fates of approximately 1,800 individual *M. sylvestris* seedlings were recorded, and of these 98% were certainly spread through grazing and browsing by cattle or horses (Buttenschon and Buttenschon 1998) after passage through the gut (Janzen 1982). Of these seedlings, some 20% may survive for longer periods if grazing ceases, whereas very few do so under continuous grazing. Almost without exception, the young shoots of Rosaceae in general are prime targets, in spite of their common thorny armament, for almost every grazing or browsing animal.

Does the dispersal of *Malus sylvestris* reveal anything about the evolutionary history of *M. pumila*? Given the very different nature of the geography, geology, and faunal pressures of northwestern Europe versus those of Inner and Central Asia, it seems likely not. In a number of ways, *M. sylvestris* illustrates the typical biology of almost all the small-fruited *Malus* species of the northern temperate zone, but not wild or cultivated *M. pumila*. Most *Malus* species of China, eastern and western Europe, and North America are small-fruited and mostly (but not entirely) bird-dispersed. Within each species, fruit size and appearance as well as tree habit are relatively uniform. Trees are predominantly solitary, rarely if ever associating in groves or anything that resembles a forest.

The idea that bird dispersal antedated mammal dispersal in apples is supported by parallels in other plant groups in various kinds of habitats where a range of dispersal mechanism is found within a single genus. For example, dispersal of species of the

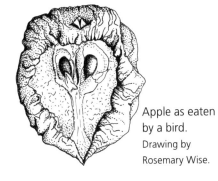

Apple as eaten by a bird. Drawing by Rosemary Wise.

genus *Aglaia,* of the mahogany family (Meliaceae), in Southeast Asia may involve fish, gibbons, civets, and birds (Pannell and Kozioł 1987), and a broad range of dispersal mechanisms is found in *Andira,* of the bean family (Leguminosae), in tropical America (Pennington 1996). Like pollination mechanisms, modes of dispersal seem to have changed and sometimes even reversed in recent evolutionary time—in terms of natural selection, such transitions seem to be readily achieved.

Like *Malus sylvestris,* the apples *M. trilobata, M. florentina,* and *"M. orientalis"* are rare inhabitants of the ancient, pre-agricultural, temperate forests of central and eastern Europe and the Middle East. *Malus trilobata* is scattered, infrequent, and generally solitary, from Palestine north through parts of Turkey into the Balkans and Greece. *Malus florentina* is found in central and southern Italy and more rarely and singly in Greece but occurs in neither the extreme north nor south of mainland Italy, nor in Sicily. It is as rare as *M. trilobata* and has been considered to be a hybrid between *M. sylvestris* and *Sorbus torminalis* (Browicz 1970) and has also been cataloged as *Crataegus florentina* or *Sorbus florentina.* Evidence from both chloroplast and nuclear DNA sequences, however, suggests that it is not a hybrid but a true wild apple species (Robinson et al. 2001, Harris et al. 2002). *Malus orientalis* (now included in *M. sylvestris*) is a tree of somewhat greater frequency, in the Caucasus Mountains and Iran; its small, hard, bitter apples are occasionally used in alcoholic drinks. The suspicion must be that these three apples, and to a lesser extent *M. sylvestris* in the narrow sense, represent escapes from the central core of *Malus* species, either at some time in the preglacial period through the continuous band of Tertiary forest, or during an interglacial. Their taxonomic distribution in three different sections of the genus (Appendix) suggests that they migrated west at different times. Having become stranded in the Middle East, or central and western Europe, the four species apparently did not receive the sustained stimuli either of geological or biotic factors bearing down on *M. pumila* and so have now become irrelevant relics of evolution, not even mentioned in the Classical literature and perhaps doomed to fairly imminent extinction.

In marked contrast to Europe, the central and southwestern region of what is now China is rich in apple species (Map 2 and Appendix; Zhang et al. 1993, Zhou 1999). However, Euan Cox (1945) makes no mention of extensive forests of fruit trees in the early collecting of *Malus* species in these areas. The notes of Ernest Wilson (1913), who was working in Sichuan at the turn of the 19th and 20th centuries, principally in the basin of the Yuan or Red River (Map 4) but with forays across the China-Tibet border, are revealing. There is almost no mention of apples, notwithstanding the number of *Malus* species present, and certainly no mention of extensive tracts of fruit forest. It is inconceivable that such careful and well-trained observers as Cox and Wilson would have failed to appreciate something resembling a fruit forest, particularly as they noted other clusters and thickets of trees and shrubs.

Such has been the rising pressure for arable land that it would probably now be impossible to re-create not only Cox's and Wilson's journeys but also any concept of the indigenous vegetation in prehuman times. The same message is related by every other plant collector or traveler in the area, past or present (Howard-Bury 1990). David Chamberlain (pers. comm. 2000), collecting plants in southern China during the 1990s, saw no sign of extensive groups of *Malus* trees. Where present, trees of the numerous *Malus* species all seem to be solitary or occur in very small groups. Cong Peihua (pers. comm. 1999) records that *M. baccata* (including *M. mandshurica*), not uncommon near his university in the province of Liaoning in northeastern China, is found singly or as very small groups, never as groves. *Malus formosana* (included in *M. doumeri*) is said to be very rare and found in mountain woodlands in Taiwan between 1,600 and 2,600 m (5,200–8,500 feet; Huckins 1972). Many of the North American species of the genus seem to follow the same pattern, though Sargent (1922) noted that the Oregon crab (*M. fusca*) and sweet crab (*M. coronaria*) occasionally occur in thickets. It remains something of a mystery as to why *M. pumila* is often so different in this respect, forming rich fruit forests in the Tian Shan.

The currently isolated pockets of *Malus pumila* in parts of the Tian Shan probably represent islands left after the retreat and fragmentation of the Tertiary temperate forest. Yet even within these small islands, and between them, there is immense variation not only in the fruit (for example, size) but also in the flower (petal color), growth habit (small shrub versus tall tree), and armament (thorns on young branches or not), as remarked upon by nearly every traveler through the region. Yet at the fundamental botanical level, due to small but significant details of the calyx and corolla, they are all one species. In its extraordinary diversity, *M. pumila* appears to differ in this respect from virtually every other species of *Malus*, which tend to be relatively uniform in growth habit, flower, and fruit detail.

Inevitably, this polymorphism, particularly of the fruit and general growth form, has resulted in the tendency of certain narrow-minded taxonomists to name new species or subspecies. Thus from the 1920s almost to the present, new names have tumbled into the literature, for example, *Malus persicifolia*, *M. kudrjashevii*, *M. jarmolenkoi*, *M. hissarica*, and *M. tianschanica*, to name only a few. None of these names has any scientific value so far as can be seen (compare Brandenburg 1991, Stevens 1991), but later in this book, attention is drawn to the quite distinctive and cohabiting *M. pumila* 'Niedzwetzkyana' and *M. kirghisorum*.

Apple country

The genus *Malus* comprises about 40 extant species (Appendix; Likhonos 1974), though as few as 8 and as many as 79 have been recognized (Ponomarenko 1986). Part of the problem of *Malus* taxonomy is the intimate and very long association that humans have had with apples. Over several thousand years the distinction between wild and cultivated species has become hopelessly blurred, hence the recognition of distinct categories is difficult. Moreover, the term "wild apple" has been applied imprecisely to all *Malus* species other than the domesticated apple. For example, in Europe, "wild apple" or "crab apple" refers to *M. sylvestris* (Plate 3), whereas in North America the terms have been applied indiscriminately to a range of species, including *M. angustifolia* and *M. coronaria*. This is further complicated by the application of the name "wild apple" to hybrids between the domesticated apple and other members of the genus *Malus*. An attempt to disentangle the meaning of the word "crab" is made in Chapter 7.

It seems likely that something resembling a species in the modern genus *Malus* arose in the middle or late Tertiary (Table 1), somewhere in the region of what are now three provinces of southern China, namely, Guizhou, Sichuan, and Yunnan (Map 2). Today, this whole region is remarkably rich in Rosaceae (Zhou 1999). Some 19 genera of the family are native to northwestern Yunnan alone. In particular, the Hengduan Mountains of northwestern Yunnan are home to 13 *Malus* species. Significantly, this whole broad area, the so-called Chuan-Dian paleoland (Zhang et al. 1993), has not been affected by recent glaciations.

Between them, these three Chinese provinces account for two-thirds of the wild *Malus* species in eastern Asia and some three-fifths of the species in the whole world. Possibly, in the preglacial period, during the interglacials, and in the immediate postglacial period, many other species spread out centrifugally from this rich center throughout the northern hemisphere. Most, for example, the species of central and western Europe, are relatively rare. Some, namely the nine native to North America (Appendix; Sargent 1922, Phipps et al. 1990), almost certainly played no part in the evolutionary history of the domesticated sweet apple. Some of these, however, now play a limited role in the breeding of ornamental cultivars (Chapter 7; Fiala 1994). By their distributions and affinities with the core species of *Malus*, it is possible that these North American species may eventually give us clues as to the periods of migration of apple species throughout the northern hemisphere.

For considerable periods of time in the Miocene (Table 1), and possibly before and since, the land bridge joining northeastern Asia with North America, through what is now the Bering Strait, seems to have been suitable for exchanges of temperate deciduous plants, and of animals as well. The great water mass of the Pacific to the south would have kept the glacial front at bay, at least at the southern edge of Beringia, whereas only a little to the east the ice descended far into North America. Thus probably via Beringia, though it may be prudent to reserve commitment as to which direction any particular migration took (Donoghue et al. 2001), many genera of familiar shrubs or small trees, for example, *Acer* (maples), *Hamamelis* (witch hazels), *Liriodendron, Magnolia,* and *Malus,* have numerous representatives on the Asian continent and generally rather smaller numbers of relatives in North America.

Such relationships, of similar relatives on different continents, are particularly marked in the so-called Appalachian forest flora (Boufford and Spongberg 1983, Wen 1999). The genera, called disjunct because they comprise geographically widely separated species, and of which there may be at least 65, are made up principally of tree and shrub species, though a few are perennial herbs. The ebb and flow of species appears to have continued until about 3.5 million years ago,

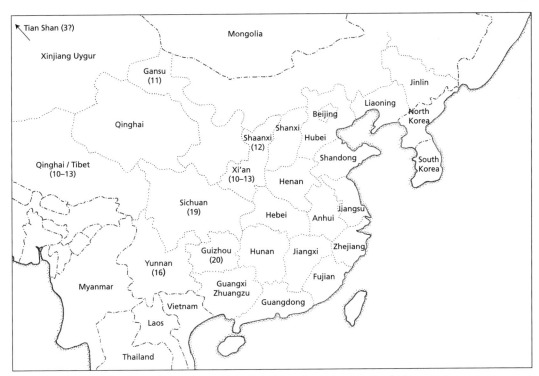

MAP 2. Numbers of species of the genus *Malus* in selected provinces only of China. Data principally from Zhang et al. (1993).

toward the end of the Pliocene and near the beginning of the Pleistocene glacial epoch. It can therefore be speculated that temperate woodland rich in species of both plants and animals enjoyed conditions that were both more widespread and farther north, probably in a continuous corridor, than they are today, though it must be borne in mind that present-day continental outlines, with the current minimal areas of glaciation, are almost meaningless in this context.

It seems highly likely that from time to time a continuous corridor of temperate forest extended from the Atlantic Ocean east through western, central, and eastern Europe to Asia, reaching North America through Beringia. To the north, the temperate forest would have graded into coniferous forest and eventually into pure taiga, the belt of coniferous forest that almost entirely circles the globe in northern latitudes. To the south, the temperate corridor would have graded successively into subtropical and tropical forests (Map 3).

Only tiny fragments of this once great forest corridor exist now in parts of Poland and Belarus. The most spectacular region is probably the Białowieza National Park, which spans the border of those two countries. In it can be found the European bison, now limited in distributed and protected. It is probable that the original bison herds from the preglacial period died out in the 1920s but once ranged throughout most of the corridor. *Cedrus* is an example of a genus of trees from this ancient corridor, from the Atlantic cedar (*C. atlantica*) in the west to the deodar (*C. deodara*) in the east. These cedars are all now generally recognized as distinct species, but they are still capable of hybridizing with one another such that some authorities consider them merely different races or subspecies of *C. libani*, the cedar of Lebanon. The remaining fragments of forest on the lower slopes of the Tian Shan of Inner and Central Asia should probably also be considered remnants of this ancient corridor.

This great continuous corridor of forest almost certainly suffered periodic fragmentation and recoalescence during oscillations of climate and geological uplift. These fluctuations might explain not only the localized distributions of rare western apple species, for example, *Malus florentina* (Browicz 1970), *M. sylvestris* (including *M. orientalis*), and *M. trilobata*, but also those of bird- and mammal-distributed apples such as *M. kirghisorum*. Periodic changes such as the ebb and flow of wetter and warmer conditions may also provide an explanation for the taxonomically ill-defined sweet apples with small fruits such as those near the lake Qinghai, formerly called Koko Nor (Map 6), south of the western part of the Great Wall (Migot 1957). Small and occasionally sweet apples (therefore not "*M. orientalis*," which is very bitter), the so-called Paradise apples, are also reported to be widespread in the Caucasus Mountains and even as far north as Kursk and Voronezh in Russian Europe, but it is not clear to what extent these are just prehistoric isolations or historic-period feral introductions. However, the idea that they all represent relics of the once widespread, preglacial, temperate forest corridor should not be ruled out.

But it transpires that only in the great ranges of the Tian Shan of Xinjiang Uygur of western China to Uzbekistan were all the conditions—geological, meteorological, tectonic, pedological, and ecological, plus very long periods of relatively undisturbed time—consistently present to drive forward the evolution of the large sweet apple of the supermarket shelf from its tiny, bird-distributed ancestors.

Inner and Central Asia can be divided into three major physical geographical regions. It is usual, now, in current geographical language, to reserve the term Inner Asia for what were the old semi-autonomous regions of the Union of Soviet Socialist Republics, now the independent republics of Kazakhstan, Uzbekistan, Turkmenistan, Kyrgyzstan, and Tajikistan, and to use the term Central Asia for those regions from the autonomous region of Xinjiang Uygur in northwestern China to the east. These three regions are the eastern monsoon region, the north-

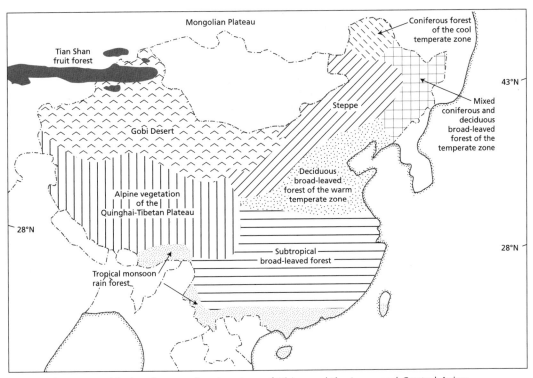

MAP 3. The geographical and vegetational regions of China and the Inner and Central Asian republics before the advent of agriculture. The continuous belt of deciduous forest was broken to the west, partly by desertification (the Gobi) and partly by the geological lifting of the Qinghai-Tibetan Plateau above the tree line.

western arid region (but including, somewhat paradoxically, the forested mountain slopes of the Tian Shan), and the Qinghai-Tibetan Plateau region, including the Himalaya with Mount Everest its highest point (Map 4).

The Qinghai-Tibetan Plateau

In the late Cretaceous the Indian subcontinent became detached from the East African coast and moved on its own tectonic plate northward at about 10 cm (4 inches) per year with respect to Asia, with which it was to collide. On the way it left behind, as geological ghosts, island groups such as the Seychelles. India collided with Asia in the mid-Eocene, about 45 million years ago (Table 1). Since then India has been forced into the great northern landmass, still being driven forward at a rate of about 5 cm (2 inches) per year. As can be mimicked on a tiny scale by pushing on the edge of a loose carpet, the effect has been that vast regions such as the Himalayan mountain range and the Qinghai-Tibetan and Mongolian Plateaux have been uplifted. The catastrophic earthquakes in Pakistan and Kashmir in September 2005 are the direct result of this pressure. The Qinghai-Tibetan Plateau has risen by about 5 km (16,400 feet; Dewey et al. 1988). In simple terms, at some point the whole subcrustal base of the plateau fell away, allowing this vast region to continue to rise like a huge table. It is now mostly some 4,000 m (13,000 feet) above sea level. In succession, as the pressure continued, the Himalaya, the Pamirs, and in their turn the great ridges of the Tian Shan began to rise. It is estimated that the Tian Shan became a distinct mountainous feature about 10 million to 12 million years ago or possibly a little earlier.

The Qinghai-Tibetan and Mongolian Plateaux have mostly risen above the tree line and, like the unvegetated talus and screes of the mountain ranges, are relatively arid because of their height. Because they are tablelands they do not enjoy the benefit of summer runoff from the high snowpacks. In addition, they have been steadily eroded by the seasonal monsoons. But of the three regions, only the eastern monsoon region and the northwestern arid region, which includes the Tian Shan, figure directly in the origin of the apple.

The monsoon region

The eastern monsoon region, principally tropical monsoon rain forest grading into subtropical evergreen broad-leaved forest and on into deciduous broad-leaved forest of the temperate zone, comprises some 47% of Central Asia.

The monsoon, powered by the warm waters of the Indian Ocean, is crucial to the early part of the apple story and still has a very significant effect in the region. The monsoon influence extends as far north as the Mongolian Plateau. In this huge monsoon zone there are considerable changes in wind direction throughout the year, and precipitation varies from season to season: "The Yunnan people say that four seasons can be found simultaneously along the same mountain slope, and different weather conditions can be experienced over a distance of 5 km [3 miles]" (Zhang et al. 1993).

Forest was the predominant vegetation before the area was settled by human beings, and the remainder was mainly desert and steppe. From the early Tertiary to near its close, these forests may have stretched in a continuous band, as discussed before, from Europe to Asia and over the Bering Strait into North America. They were then fragmented in the west by the glaciers, and in Asia by the advancing deserts (Table 1).

The soft, fertile soils of the eastern monsoon region are derived, in part, from massive deposits of loess. The word *loess* (German for dust) was coined in the 1870s by Ferdinand Paul Wilhelm von Richthofen; it is defined as windblown

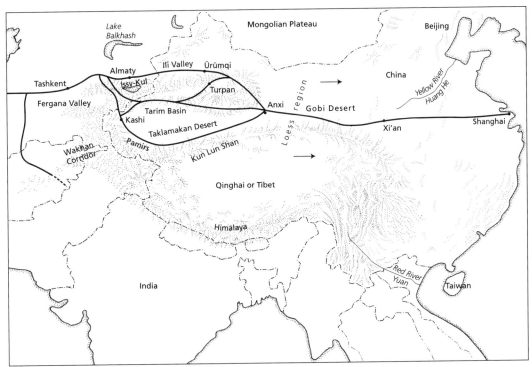

MAP 4. Loess blows from the uplifted Mongolian and Qinghai-Tibetan Plateau, principally eastward into central and eastern China.

dust, rich in silica, with a grain diameter of less than 0.06 mm. The loess blows off the high Mongolian and Qinghai-Tibetan plains and the Gobi Desert for many months of the year (Map 4); for as long as 2.75 million years it has coated this region with its deposits. Loess deposits cover about 440,000 km² (170,000 square miles) in China and in places, for example, near Lanzhou on the Huang He or Yellow River, are more than 300 m (980 feet) thick. Most of the surface of this eastern monsoon region lies below 1,000 m (3,300 feet) in elevation, and much of the eastern part even lower, below 500 m (1,640 feet), with many low-lying alluvial and loess-coated plains. The combination of loess and high summer rainfall of the area renders it well suited for agriculture.

This whole huge area, of what was once tropical and temperate forest, was described, and its devastation regretted, by the great plant collector Ernest Wilson, for even by the early 20th century it had been extensively denuded of its wild vegetation. As Reginald Farrer, in typical flowery style, wrote in his *On the Eaves of the World* (1926), describing his travels in 1914, "The face of Southern and Central Kansu [Gansu] is one vast corrugation of those tedious loess downs, till the hunter after wild mountain-lands grows every day more sick with the inexhaustible energy of the Chinese cultivators that has thus bared all the world for his needs and left no standing room anywhere for wild beast or tree."

The whole vast region, protected by the warm air and water of the Indian monsoons, has remained essentially ice-free in both the distant and immediate past except for local glacial spreads in the Himalaya and other mountainous regions. Northern Europe and North America, on the other hand, with no sufficiently large water masses of stored heat to the south to protect them from the cold, have been repeatedly ice-scraped up until about 12,000–10,000 years ago. The existence of great warm water masses to the south impeded the advance of the ice into Central Asia. Moreover, in the Cretaceous the Indian subcontinent was positioned off the eastern coast of Africa, leaving an even larger warm sea area to the south of the Asian landmass. Likewise, the great sea mass of the northern Pacific would have held back the ice and permitted, from time to time, a habitable land bridge, Beringia, to form between Asia and North America. Thus, in a seeming paradox, the locking up of vast masses of water in ice joined, on more than one occasion, eastern Asia and North America. There were also, in this last glacial onset, about 20 warmer interglacial periods, the last being today's, beginning about 12,000–10,000 years ago.

Since the most recent glacial period, the floras and faunas of Europe and North America have slowly recolonized the ice-scraped zones (Willis 1996). In contrast, despite the geological turmoil, or perhaps as a result of it, the eastern monsoon region of southwestern China into Inner and Central Asia became un-til the arrival of destructive humankind a sanctuary and breeding ground for a wide and constantly diversifying range of animals and many thousands of plant species. In broad terms, this area today has 10 times the number of plant species of western Europe and 30 times that of the British Isles. It is thus no surprise to realize that the so-called English garden is, in major part, a product of plant col-lectors' diligence, particularly over what is now principally Inner Asia and the rest of China, and owes little or nothing at all to the indigenous, impoverished, postglacial northern European flora.

The Tian Shan

The northwestern arid region comprises several distinct desiccated areas, but the major feature germane to the story of the apple is the Tian Shan, in translation the heavenly or celestial mountain range. The great range of the Tian Shan rose steadily out of the Central Asian landmass. Then, from about 2.75 million years ago, in the late Pliocene or possibly a little earlier, deserts began to form (Table 1). The Tian Shan stretches just over 1,600 km (1,000 miles) from the Xinjiang Uygur autonomous region of western China to Uzbekistan in the west, passing through Kazakhstan and including Kyrgyzstan and Tajikistan. In the east, the Tian Shan lies between the dry Tarim and Junggar basins and comprises more than 20 parallel ridges and valleys running roughly in an east–west direction. This predominantly east–west orientation is significant to our hypothesis of the origin of the sweet apple.

The Tian Shan does not belie its name, the heavenly mountains (Plate 14). With its jagged, glistening, snow-covered peaks, forest-clad slopes, and high, sheltered pastures bejeweled in spring with flowering bulbs and fruit blossoms, and in the autumn with a cornucopia of fruit, the Tian Shan is the apotheosis of a favored, ancient mountain kingdom. In Alexander Borodin's opera *Prince Igor* (about 1870), the Polovtsian dancers, slave girls from Kazakhstan, sing with pas-sion and sorrow of their homeland, "Fly away, my song, on the wings of the wind, to my dear motherland, where the crimson roses bloom in the valleys and night-ingales sing sweet in sunlit pastures." It would seem that Borodin, who himself had never been nearer to Kazakhstan than Moscow, had talked to Kazakh girls taken back to the capital by officers of the Imperial Russian army. He had bor-rowed and adapted both their folk-song theme and memories of their distant homeland for his music. Every favored visitor to their former pastures can empa-thize with them in their misery.

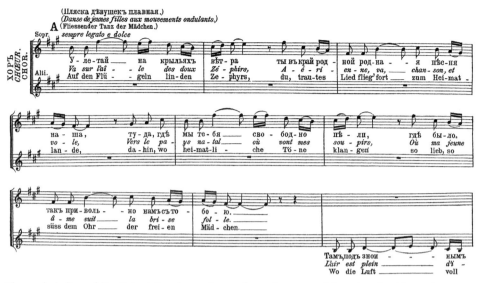

The melody line of the Polovtsian dance theme from the opera *Prince Igor* by Alexander Borodin.

Slowly, the deserts, which have been given many names, began to close off the Tian Shan in almost all directions. A general name for the eastern deserts is the Gobi. The word *gobi* is not the proper name of a geographical area but rather an expression commonly used by the Mongols to designate a particular association of geographical features: wide, shallow basins of which the smooth bottom is filled with sand, pebbles, or more often, gravel. The Gobi measures nearly 2,000 km (1,200 miles) from north to south near the 104° meridian and follows the 44° north parallel. It reaches east toward the western Khingan Mountains; to the west its extent is only limited by the use of the word *gobi*. It lies in an uninterrupted stretch over the wilderness of the Junggar basin and the wastes of Xinjiang Uygur, separated from each other only by the hilly and fertile belt of the Tian Shan. Thus from the Pamirs of Tajikistan in the west to the borders of Manchuria or Heilongjiang province in the east, the Gobi extends some 5,760 km (3,600 miles).

A most remarkable feature of the Tian Shan is the patchy and heterogeneous nature of the landscape, with very different plants, and presumably animals, across it. More than 200 rivers, of which one of the most spectacular is the Ili, originate in the range, but all run into internal drainage basins. Most of the ridges are more than 3,000–4,000 m (9,800–13,000 feet) above sea level, towering over the deserts to the north, east, and south. To the west, the mountain slopes grade away through Uzbekistan into the desert and steppes of Turkmenistan. On the northern slopes, protected somewhat from the fierce, unrelenting summer sun, average annual precipitation may be as much as 400–500 mm (16–20 inches). Here and there, rivers, lakes, and shallow underground streams augment this to the benefit of nearby plant populations. These local refuges may influence the sporadic distribution of the apple species central to our hypothesis, *Malus pumila,* and its strikingly variable abundance.

Mountain uplift and massive tectonic fault movements, which never cease in the whole Tian Shan region, constantly expose a range of variably weatherable rocks through erosion (Plate 16). Some immediately form alluvial fans; others, the harder rocks, persist as ridges or peaks. Many parts of the range rise by as much as 1.5 cm (⅗ inch) per year. The whole region is crisscrossed with a myriad of fault lines, almost all of which are active (Map 5). Adding to the diversity of rock and soil chemistry was the draining of the remnants of the Jurassic Tethys Sea, which divided the ancient supercontinent of Pangaea and stretched from what is now Britain to Central Asia, leaving local deposits of calcium carbonate and salt (Plate 18).

Therefore, for a considerable time up to today, the land surface of the region north and west of the Indian thrust, including that of the Tian Shan, would have been twisted, distorted, and broken, all the time bringing ancient rocks of every geological facies to the surface, creating fresh canyons, cliffs, and caves, altering drainage patterns, exposing fresh soil, and breaking established vegetation cover. The Mongolian and Qinghai-Tibetan Plateaux, in contrast, as explained before, were lifted up more or less complete as tables and suffered little or no such twisting and breaking, hence no such soil invigoration.

In contrast, the lower slopes of the mountain ranges, particularly those of the Tian Shan, have received constant soil reinvigoration; they have remained very fertile, at least locally, with the added advantage of a very long glacier-free history. Ungraded by glaciers, the dazzling peaks of the Tian Shan, in particular, have retained their jagged outlines. Indeed, they resemble the mountains a child might draw. Elsewhere, glacial sheets began to descend across northern and western Europe and North America from about 7 million years ago (Table 1). Glaciation was instrumental in the destruction, movement, and spread of many animals and plants in Asia and, across what is now the Bering Strait, to and from North America.

In the ancient and multilayered deposits of the Tian Shan, internal and external streams, which run all year and are fed by the extensive snowpacks, have washed out great channeled labyrinths. As the rocks break apart under the stress of tectonic movements, the internal lacunae are exposed as caves and tunnels, often yielding the appearance of a geological Gruyère cheese. Cave systems become of significance in the occupation of the ranges by bears (Chapter 2). Erosion patterns leave, on the one hand, vertical cliffs of hard white limestone or, on the other, tumbling scree slopes of soft sand or unconsolidated rock debris, which are slowly colonized by forest vegetation (Plate 16). Landslides were, and are, frequent. Within the memory of people living there, natural mountain lakes burst and inundated thousands of square kilometers, washing away whole cities, agricultural land, and vegetation, creating floodplains for new development. By the standards of, say, geologically quiescent Britain, it is not a comfortable region, but at every turn, and in every season, is visually dazzling and dynamic.

The faunas and floras of many regions of the world, for the reasons set out here, are spectacularly depauperate. The shapes of the continents and the barriers to migration seem largely to be responsible for surprising anomalies in distributions of organisms. In western Europe, the great Scandinavian ice sheet, advancing toward the Alpine massif, almost completely pinched out plant life, with the exception of what are now southwestern Ireland, southwestern Spain, southern Portugal, the southern Balkans, and southern Greece, which together constitute a huge westward protrusion of the Eurasiatic plant world. Barred from further migration south by the Mediterranean, many temperate and warmth-loving species became extinct in Europe and, except by human reintroduction, have never been able to recolonize the area.

The same pattern was generally true of the North American continent since virtually no food crops survived there. Hence, it was not possible for the earliest human colonizers there to domesticate horses because there were no indigenous oats, peas, or beans to feed them in winter.

But in eastern Asia, with increasing northern cold, the Tertiary temperate and tender plants were able to retreat south or spread away from locally developing mountain ice packs without serious restriction and to recolonize as conditions improved. Crop plants, or their ancestors, were still present in the Kazakh–Uzbek area, hence the stimulus to domesticate equines there. At the same time, the powerful monsoon influence kept at bay, to an extent, the continentality of climate. Thus Inner and eastern Asia, over many millions of years in the key period for flowering plant evolution, that is, from about 100 million years ago, were able to retain and develop a richness of flora denied to western Europe or North America.

A glacial advance locks up very large quantities of water in the ice sheet, and glacial regression has the reverse effect. Such events make nonsense of the generally accepted continental outlines, and the extent or even the existence of inland seas or great lakes. It is estimated that in the glacial cycle, sea levels may have changed by as much as 150 m (490 feet).

Moreover, it would not be true to assume that with the onset of glaciation at the boundary between the Tertiary and Quaternary periods, glaciers were universally present. The Quaternary is characterized by oscillations in the extent of the land ice in the northern hemisphere. There seem to have been perhaps 20 brief interglacials, one about every 100,000–200,000 years. The latest 160,000 years of the Quaternary comprise the last full cycle of glacial advance and current retreat. For example, there seems to have been a particularly warm period in the Pleistocene about 120,000 years ago.

Although the great area of Inner and Central Asia has progressively desiccated since the end of the last glaciation, there have been at least brief wetter intermissions apart from the warm periods. From an analysis of the bacteria-induced changes in manganese content of desert varnish, it seems that the Gobi may have entered two wet periods, beginning about 3,000 and 8,000 years ago (Broecker and Liu 2001). These fluctuations, forcing elevational changes in the fruit forest of the Tian Shan, helped drive the evolution of the sweet apple.

CHAPTER 2

Origin of the apple

NIKOLAY VAVILOV argued that the sweet apple evolved in Kazakhstan, where populations of wild apples, formerly called *Malus sieversii*, are extremely variable in tree form, flower color, and fruit size, color, and flavor. The wild *M. pumila* probably evolved from a bird-dispersed species resembling the modern-day Siberian crab, *M. baccata*. DNA evidence strongly supports Vavilov's ideas and makes dubious any role of hybridization between wild *M. pumila* and any other species, including the European *M. sylvestris*, which has often been implicated.

Three hypotheses are offered to explain how the wild apple got to the fruit forest of the Tian Shan, where it now grows in its greatest variety. The principal dispersal agents there turn out to be bears and horses.

WITH REMARKABLE FORESIGHT, Nikolay Vavilov (1887–1943) suggested that the wild apple of Turkistan (comprising modern-day Turkmenistan, Uzbekistan, Kyrgyzstan, Tajikistan, southern Kazakhstan, western China, and northeastern Afghanistan) and its close relatives were the ancestors of the domesticated apple. He traced the whole process of apple domestication to Almaty in Kazakhstan. Since the wild apple there bore fruits similar to those of the domesticated apple, Vavilov reasoned that it must have been one of the major progenitors. There was much speculation and few real facts.

More recent fieldwork in the region (Forsline et al. 1994, 2003, Forsline 1995, Soest et al. 1998, Juniper et al. 1999, Luby et al. 2001, Robinson et al. 2001, Harris et al. 2002) appears to confirm the very close similarity between the wild apple (what he called *Malus sieversii*) and cultivated apples suggested by Vavilov, though Vavilov himself hinted at subsequent hybridization. Furthermore, "this area, what we now recognize as Inner and part of Central Asia, is the area of greatest diversity and the center of origin" of the domesticated apple

(Janick et al. 1996, Janick 2003), which is linked to Vavilov's somewhat oversimplified and much criticized idea that the center of species diversity is its place of origin (Harlan 1992). In this case, however, it looks as though Vavilov's early speculation was probably almost completely right. Nevertheless, we should recognize that the richest area of species diversity in the genus *Malus* is in central and southern China (Map 2).

Malus pumila and its relatives

The Inner and Central Asian wild apple is an astonishingly diverse species with a wide range of tree habits, flower colors, and fruit forms, colors, and flavors (Way et al. 1990, Zhou 1999, Geibel et al. 2000, Fayers 2002: fig. 3/6). In addition, its genetic diversity, as measured by differences between individual plants in the structure of their enzymes, is significantly greater than that in four widely distributed North American wild apples (Dickson et al. 1991, Lamboy et al. 1996)

Malus pumila is closely related to two apple species that have smaller, more uniform fruits and differ from *M. pumila* in that they seem to possess a normal endosperm rather than having almost none. *Malus baccata*, the Siberian crab, has small red fruits that generally ripen in early September, hang in clusters, and have seeds that are bird-dispersed. Bird-distributed fruits generally hang on long, flexible pedicels, allowing them to move in the wind; birds are particularly well adapted to sense such small movements. But *M. pumila* and its cultivars hang on stiff, short pedicels, usually unshaken by even the strongest gale. *Malus kirghisorum* (Plate 4) appears to be intermediate between *M. baccata* (Plate 5) and *M. pumila* (Plates 1 and 2), perhaps reflecting a less intense period of selection in the *M. pumila* direction. It seems possible that an ancestor of an apple similar to *M. baccata* may have had a wider distribution than nowadays and that populations of one or possibly more of these bird-dispersed apples might have become trapped as the Tian Shan, thrust upward by the northward-migrating Indian subcontinent, began to rise out of the ancient Tethys Sea.

In terms of domestication, the species of section *Malus* are beyond dispute the most important in the understanding of the origin of the sweet apple (Appendix). But the taxonomy of the species in section *Malus* is complex. With few discrete morphological or anatomical features that have clearly precise boundaries, the difficulty of species delimitation has hampered investigations of apple origins (Rohrer et al. 1994). Such problems contributed to the statement that "it is therefore futile to try to delimit the area of initial domestication on the basis of the evidence available from the living plant" (Zohary and Hopf 2000). However, the revolution in molecular systematics provides an excellent source of consistent and clear characters, if used with discretion, for taxonomic analysis.

DNA evidence

In large part, genetics, whether of the classical or molecular kind, depends upon mistakes: mutations. Were the replication of the genetic message perfect in every way in space and time, evolution would not happen. Mutations are one of the major engines of evolution, so they can reveal relationships and be used as a method of identification. DNA fingerprinting, one technique among many now available to the molecular geneticist, can determine to a very high degree of probability whether or not a suspect was present at the scene of a crime, for example.

In plants, there are three sources of DNA within every cell, all of which can be used to detect relationships through their mutations. Individually or in combination, nuclear DNA (nDNA) and cytoplasmic DNA, which includes chloroplast DNA (cpDNA) and mitochondrial DNA (mtDNA), bear molecular markers that are passed from parent to offspring, providing important clues about the origins of domesticated plants, for example, beans (Becerra Velásquez and Gepts 1994), potatoes (Hosaka 1995), cassava (Olsen and Schaal 1999), and *Citrus* (Moore 2001, Mabberley 2004).

The cpDNA provides data about evolutionary relationships of the maternal line because chloroplasts are transmitted to the next generation through eggs (in the ovules) but not sperm (from the pollen). In *Malus*, the chloroplasts and, incidentally, the mitochondria are exclusively transmitted down the female line, though this is not necessarily true in all plants (Ishikawa et al. 1992). The nDNA, which is inherited from both parents in more or less equal amounts, provides independent data, which combined with cpDNA may shed light on the origin of domesticated apples. It had been expected that hybridization and introgression (the flow of genes between genetically different groups through hybridization and backcrossing) had been factors in the development of domesticated apple cultivars. Therefore, for investigations that aimed to determine the earliest origins of the domesticated apple, it is important to use cultivars that are as ancient as possible.

In a survey that sampled apples across the entire genus *Malus*, *mat*K, a cpDNA-encoded region about 1,800 base pairs long (of which 1,341 base pairs were sequenced), only 16 phylogenetically informative characters, that is, mistakes in the code that can suggest lines of evolution, were found. The analogy is the message passed along the trench: "Send reinforcements, we are going to advance," which arrives as "Send three and fourpence, we are going to a dance." It is likely that, from the construction of this sentence and the probability of error in the English language, the direction of inheritance was in the manner indicated. Likewise, knowledge of the probability of error and its direction in the base pairs can provide information on the direction of evolution and, to a limited extent, on its timing.

There was poor resolution, that is, not unequivocal evidence of the direction of evolution, in the phylogenetic tree (Lamboy et al. 1996). However, 39 base pairs from the 3′ end (the end of the sequence from which the code is read) of the *mat*K coding region, two duplications were found (Robinson et al. 2001, Harris et al. 2002). A duplication is a region of the gene that is replicated either accurately (a perfect duplication) or inaccurately (an imperfect duplication). Duplications, whether perfect or imperfect, can be valuable in the elucidation of evolutionary trees. Duplication 1 here is an imperfect 8-base-pair-long duplication that differs by a thymine residue, and it is found in a number of *Malus* species from sections *Malus* and *Sorbomalus*. Duplication 2 is perfect and 18 base pairs long; it occurs in a number of forms in both the Inner and Central Asian wild apple and the domesticated apple, indicating more than one introduction and thus suggesting that the Asian wild apple may be the major maternal contributor to the domesticated apple (Watkins 1995).

Analysis of ribosomes from the nucleus, specifically that DNA region called the internal transcribed spacer (ITS), yielded 89 phylogenetically informative characters from the 617 base pairs sequenced. Thus almost 15% of the apple ITS contains mistakes that could prove useful in elucidating ancestry. A strongly supported group, one in which the evidence is unequivocal, comprises the Inner and Central Asian wild apple and the domesticated apple (including 'Niedzwetzkyana', Plate 12) plus *Malus asiatica, M. orientalis* (now a synonym of *M. sylvestris*), and *M. prunifolia* (all in section *Malus,* series *Malus;* Appendix). Together, data from the chloroplast and nucleus support the view that the domesticated apple is most closely related to the species of series *Malus.* Additional phylogenetically useful markers need to be sought, and additional sampling of wild and domesticated apples is needed to ensure that rare hybridization events, which may have played a role in the early domestication of apples, have not been overlooked.

The desire for additional DNA markers has led to the use of other molecular techniques, including use of markers from randomly amplified polymorphic DNA (RAPDs; Dunemann et al. 1994, Guilford et al. 1997, Jones et al. 1997, Zhou and Li 2000) and nuclear microsatellites (SSRs; Hokanson et al. 2001, Wünsch and Hormaza 2002). But molecular evidence is not always conclusive. Dunemann et al.'s RAPD results seem to indicate that *Malus pumila* and *M. syl-*

vestris (Plate 3) were involved in the origin of cultivated apples, but they suggested that *M. sylvestris, M. florentina,* or *M. dasyphylla* could be the female parent of *M. pumila.* In contrast, Hokanson et al.'s SSR studies did not turn out to be useful in determining relationships among the 142 accessions from 23 species analyzed. Plus, it has been argued that data from RAPDs are not appropriate for the reconstruction of phylogeny because of problems involving reproducibility, primer structure, dominance, product competition, homology, allelic variation, genome sampling, and nonindependence of loci (Harris 1999), and similar problems have been highlighted for the use of SSRs (Robinson and Harris 2000).

Evidence from cultivars

Many thousands of apple cultivars have been selected for use as desserts, in cooking, and for making cider and have subsequently been propagated vegetatively for hundreds of years in Europe, Asia, North America, and more recently in the southern hemisphere. In the British Isles there are thought to be some 2,500 distinct cultivars, with currently 2,138 at Brogdale in Kent alone (Smith 1971, Janes 1998). A similar number may exist in the United States and Canada; probably more than 2,500 are now assembled at Geneva, New York, near Cornell University (Way 1976, Forsline et al. 2003). In what was the territory of the former Soviet Union there are possibly as many as 6,000 cultivars. It is likely that there are as many as 20,000 named cultivars in the world. Most of these, together with wild species, are maintained in national collections as genetic resources for breeding, particularly as sources of resistance to apple scab (caused by the fungus *Venturia inaequalis;* Manganaris et al. 1994), powdery mildew (*Podosphaera leucotricha*), and bacterial fire blight (*Erwinia amylovora*). Extensive programs, both by conventional plant breeding methods and by genetic modification, are underway to attempt to counter these afflictions (Geibel et al. 2000, Lespinasse and Aldwinckle 2000, Norelli et al. 2000, Luby et al. 2001, Ko et al. 2002, Ferree and Warrington 2003).

Fate has not been kind to ancient established orchards or apple collections. In World War I the famous nursery and large tree collection of Simon Louis-Frères (1896) near Metz, France, where many ancient cultivars were being grown and from which many apple cultivars and other fruits and ornamentals had been derived, became the front line between the opposing Allied and German forces and were destroyed. In World War II the last great tank battle of the conflict was fought in July 1943 between the Red Army and the Wehrmacht through the apple-growing and fruit nursery area of the Kursk region of European Russia.

Also, anti-Mendelian repression in the Soviet Union ushered in almost two decades of persecution of both genetics and conventional geneticists. Trofim Lysenko's bizarre theories of inheritance, supported by Stalin, crippled Soviet genetics and plant breeding, beginning in the late 1930s and lasting into the 1950s. They contributed to widespread food shortages and resulted in the dismissal or even death of many geneticists—including one of the heroes of the apple story, Nikolay Vavilov (d. 1943)—the closing of genetics laboratories, and the dispersal or destruction of fruit and cereal collections (Harman 2004). Many of these collections and orchards had been established in the first half of the 20th century by Vavilov. Since then, Chairman Mao's Red Guards, in their destruction of records and libraries in the years 1966–1976 likewise did not advance horticulture in China.

As with that of the wild species, the taxonomy of apple cultivars is fraught with problems, including environmental and developmental variation, and the limited number of morphological and chemical characters available to distinguish them (Rohrer et al. 1994, Watkins 1995). In addition, the triad of increasing collection size and running costs yet decreasing budgets traps a collection manager whose goal is to have well-characterized, easily used collection, forcing choices to minimize genotypic redundancy. Thus molecular genetic markers are potentially valuable in identifying duplicates (and recognizing cultivar synonyms), correctly naming misidentified plants, and enabling managers to reduce collection size without diminishing its genetic variation (Nybom et al. 1990, Savolainen et al. 1995, Noiton and Alspach 1996, Wünsch and Hormaza 2002).

Two contrasting types of molecular genetic markers can be used to attempt to identify apple cultivars. The first type includes those that identify many low-information-content loci, such as RAPDs. Studies of variation in RAPDs in *Malus* have focused on the diversity present in domesticated apples and germplasm collections. Such variation in 52 putative loci in a collection of 27 apple cultivars was sufficient to differentiate them. RAPDs have been used to analyze the maternal and paternal contributions to pedigrees (Harada et al. 1993, Dunemann et al. 1994, Landry et al. 1994). In a large study to assess variation in 50 old domesticated apple cultivars and current breeding lines, and a collection of 105 other apples, including wild species, 43 RAPDs were used (Oraguzie et al. 2001). Individual cultivars could be distinguished, and about 95% of the RAPD variation occurred within cultivars, breeding lines, and wild species. Furthermore, Inner and Central Asian wild apple accessions were scattered in a cluster analysis based on genetic distance, suggesting that the variation seen in cultivars and breeding lines is present in the wild populations. RAPDs have been criticized for the analysis of genetic variation on both technical and theoretical grounds (Harris 1999), however, and given the problem of poor interlaboratory result reproducibility, arbitrary fragment length polymorphisms (AFLPs) may be a more effective way of analyzing genotypes using many low-information-content loci (Jones et al. 1997).

The second type of molecular marker includes those that identify a few high-information-content loci, such as SSRs (Guilford et al. 1997). For example, finger-printing was able to distinguish clearly between the cultivars 'Golden Delicious', 'Jonathan', 'Red Delicious', and 'Rome Beauty' (Nybom 1990a, b, Nybom and Schaal 1990, Nybom et al. 1990). Eight apple SSRs (seven dinucleotide-repeat loci and one trinucleotide-repeat locus) were used to identify genotypes in a core collection of 66 domesticated apple cultivars (Szewc-McFadden et al. 1995, 1996, Hokanson et al. 1998). All but 7 pairs of accessions (5 of which were either muta-tions or their progenitors) of the 2,145 possible pair-wise cultivar combinations could be distinguished. The probability of any two single-locus genotypes match-ing by chance in this study ranged from 0.027 to 0.970, which dropped to 0.156 $\times$ 10^{-8} for comparisons at multiple loci (multiple positions on a chromosome). This high discrimination power shows the value of SSRs for cultivar genotyping, and their codominance and high repeatability make them more reliable markers than RAPDs. Furthermore, SSRs appear to be very useful anchor markers in apple genome mapping projects. The occurrence of low-frequency alleles makes SSRs potentially valuable markers for pedigree analysis, though large numbers of codominant markers are likely to be necessary if complex crossing patterns are to be disentangled.

Limited surveys of apple cultivars have revealed two cpDNA mutations that together provide evidence for multiple origins of the maternal component of do-mesticated apple, which is not surprising because apple cultivars are almost en-tirely propagated vegetatively. A survey of 40 cultivars showed that they could be divided into two cpDNA haplotypes (unique combinations of genetic markers in a uniparentally inherited genome), depending on a cytosine–thymine transition at position 17 of the *atp*B–*rbc*L spacer (Savolainen et al. 1995). Similarly, a survey of 9 cultivars showed that they could be divided into two groups based on the occurrence of an 18-base-pair duplication in the *mat*K–3′ *trn*K spacer region (Robinson and Harris 2000). The duplication was found only in the domesticated apple and one accession of the Inner and Central Asian wild apple collected by B.E.J. in Uzbekistan. Unfortunately, since the objectives of these two studies were different, complementary cultivars were not included. However, combining these two markers into the same study may allow major lineages in cultivated and wild apples to be identified.

The origin of the apple: A tentative conclusion

The molecular data reveal that members of section *Malus* (Appendix), particularly the wild apple of Inner and Central Asia once known as *M. sieversii*, were overwhelmingly if not uniquely important in the origin of the domesticated sweet apple. The extraordinary and as yet incompletely understood morphological, biochemical, and molecular variation within this wild apple species appears to account for the variation seen in the domesticated sweet apple. Selections from the fruit forest of the Tian Shan, likely on more than one occasion, could be the source of all the variation seen in the domesticated apple, without the involvement of any other species though the opportunity would seem to have been available (Coart et al. 2003). The timing of flowering, at least, for other Western species would have permitted cross-pollination.

Nevertheless, it has to be admitted that sufficient markers do not yet exist to make workers completely confident that no early hybridization took place. Later hybridizations, however, for example, 'Maypole', which may have involved the eastern North American sweet crab, *Malus coronaria,* and 'Wealthy', raised in 1861 in Minnesota and which may have involved *M. baccata,* have been important in the creation, particularly in U.S. breeding programs, of a small number of new cultivars that have economically important characteristics. More detailed analysis requires much more sampling of the variation within Inner and Central Asia. Given the extensive range but broken distribution of the wild apple, analysis of old apple cultivars using highly variable molecular markers (for example, SSRs), identification of DNA sequences that permit resolution of section *Malus* species, a comprehensive taxonomic account of the entire genus, and additional archaeological evidence for human use of apples in Inner and Central Asia and the Middle East will be important in drawing a final conclusion.

When and how did the precursor or precursors of *Malus pumila* migrate into the rising Tian Shan region? Were they trapped there and modified in situ by the effects of the rising landmass consequent upon the Indian orogeny? From the strongholds of *Malus* species farther to the east and south, and perhaps spreading along the Gansu Corridor (a fertile strip that runs along the base of the mountainous Qilian Shan on the border of Qinghai and Gansu provinces, that separates the Mongolian Plateau and the Gobi from the Qinghai-Tibetan Plateau, and that points northwest toward the Tian Shan; Map 5), there may have arrived a wild apple species resembling *M. baccata*. When this happened it is impossible to say, and the colonization may possibly have been at any time from the emergence of the Tian Shan up to about 3 million years ago or even closer to the onset of glaciation.

It should not be thought that the *Malus* was the sole migrant. Today, the Tian Shan area contains the greater proportion of the temperate-zone fruits of the northern hemisphere, including species of pear (*Pyrus*), mountain ash (*Sorbus*), hawthorn (*Crataegus*), *Cotoneaster,* and quince (*Cydonia*). These often grow alongside wild species of apricot, cherry, and plum (*Prunus*) together with small-berrying plants such as bilberry and cranberry (*Vaccinium*), blackberry and raspberry (*Rubus*), gooseberry (*Ribes*), grape (*Vitis*), and strawberry (*Potentilla,* including *Fragaria*). In places, too, can be found species of elder (*Sambucus*), white mulberry (*Morus*), and sea buckthorn (*Hippophae*) along with almond (*Prunus dulcis*), pistachio (*Pistacia*), hazelnut (*Corylus*), and walnut (*Juglans*). The term "fruit forest" is therefore not ill-chosen. Little wonder, too, that now, combined

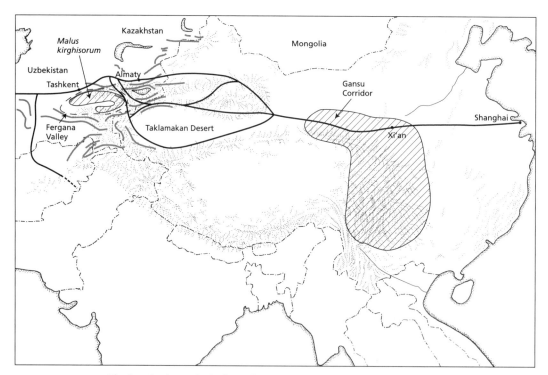

MAP 5. The hatched area including Xi'an represents the general area of high concentration of *Malus* species, with the Gansu Corridor extending to the west. The hatched area around the Fergana Valley indicates the distribution of *M. kirghisorum*. It is surrounded by areas of fruit forest, containing *M. pumila*. Original animal migration routes that later became the east–west trade routes are shown as solid lines. Note that the trade routes pass through regions with diverse apple species. Gray lines represent the major fault zones in and around the Tian Shan.

with the mountain meadow herbs, the whole length of the Tian Shan is a huge commercial honey factory (Plate 19).

When, howsoever and by what vector, early apples or their seeds may have reached the fruit forest of the Tian Shan is unknown. But three hypotheses can be put forward. The Tian Shan is at least 10–12 million years old, young by mountain-building standards—it is still rising. As explained in Chapter 1, there are areas in these mountains where a spot will be 1.5 cm (⅗ inch) higher above sea level at the end of a year than at its beginning. There is a constant state of geological revolution generated by slippage along the myriad of fault lines of the region.

Hypothesis 1. There must have existed somewhere in what is now central China, roughly in what is now the area of Xi'an, Shaanxi, perhaps in the great corridor of Tertiary temperate forest and probably more widespread than today, a recognizable ancestral apple, at least by some 10–12 million years ago. This eo-apple, which was almost certainly bird-dispersed, might already have existed, in its northern and western spread, in the more benign climate of the Tertiary forests and have reached the Tian Shan. As the Tian Shan rose, this eo-apple became detached, both in the geophysical and evolutionary sense, from contact with its Chinese, Russian, and would-be North American relations and began its unique evolution in the changing environment of geological turbulence but glacial protection that characterized the Tian Shan. This is the basic vicariance hypothesis.

Hypothesis 2. The dispersal hypothesis suggests that the Tian Shan was only secondarily invaded by one or more eo-apples, long after its distinctive and isolated geological and climatic features were established. Birds, particularly the azure-winged magpie, are dispersers of some present-day apple seeds (but probably not those of *Malus pumila*). Apple seeds carried in the birds' crops or clotted onto feet or feathers could have been deposited here and there along migration routes over the hostile separating regions into areas of suitable habitat.

Hypothesis 3. A third suggestion, but one that does not exclude the other two, is a development from hypothesis 1. As climatic changes in successive ice ages expanded and contracted the potential ranges of plants, and as the Tian Shan rose and rose, twisting and distorting its shape, the early Tian Shan apple populations were isolated into geographically separated refuges. As these disjunct populations came together, perhaps more than once, and interbred again, their offspring were subjected to selection by every aspect of the earth movements and soil variation. All the rich potential of evolution became manifest in this small region but over an enormous period of evolutionary time. This is the refugia-coalescence hypothesis.

Regardless of the possible patterns of evolution just outlined, the eo-apple, penetrating into the rising Tian Shan, perhaps via the Ili Valley (Map 4), would have encountered a continuum of habitat to the west. But to the east, even if small quantities of seed could escape over the high mountain passes, by whatever vector, there was soon no open migration route. The wild horse and wild donkey might have carried seeds in their bellies or crusted in mud on their hooves, but they would have defecated or shed this seed into the inhospitable and ever-expanding sand of the Gobi to the east, or the Taklamakan to the south, or the arid, salty Turpan depression that stretches vast distances to the north, east, and south. But to the west, the migratory door had opened and was to remain open until the arrival of human beings, and it remains open.

The fruit forest

The wild populations of *Malus pumila* were discovered in 1793 by a German-Russian botanist, Johann, or Ivan, Sievers, during his travels in the mountainous Khrebet Tarbagatay of southeastern Kazakhstan (Map 6). This area is still a refuge for some of the best of the small and now endangered areas of fruit forest. Sievers's sudden death prevented his describing this species, and it was not until 1830 that Carl Friedrich von Ledebour in his *Flora Altaica* (of a broad area including the far eastern tip of Kazakhstan and parts of Mongolia, China, and Russian Siberia) named the species *Pyrus sieversii* in his honor. Max Roemer transferred the species to the genus *Malus,* hence *M. sieversii* (Likhonos 1974).

So *Malus pumila* (which includes the naturally occurring populations once called *M. sieversii*) is a native of the mountains of Inner and Central Asia, principally the Tian Shan. It grows wild in the forests that lie as very ancient, unglaciated, isolated islands in the most humid parts of those mountains. A long period of isolation, yet subjected to heavy natural selection pressures, may provide a clue to the unique characteristics of the forest of fruit trees in which *M. pumila* is now found. This fruit forest is found as far as north as 47.30° in the Tian Shan (Forsline 1995). In places in this area the distribution of *M. pumila* must overlap with that of Siberian crab, *M. baccata;* B.E.J. has seen *M. baccata* as far south as 44.00°. *Malus pumila* also overlaps in its distribution with *M. kirghisorum.*

Few travelers with botanical knowledge passed through these regions in the early historical period. However, the Chinese traveler and statesman Ye-lü Tch'utsai (1190–1244; sometimes spelled Yeh-lü Ch'u-ts'ai or Yelü Chucai) accompanied Genghis Khan in his attempted conquest of the West. Being an embedded war correspondent with the Golden Horde must have been an interesting if somewhat hazardous assignment. But he survived and wrote a book about his travels that he called *Si-yu-lu* (sometimes spelled *Xi You Lu*), *An Account of a Journey to the West*. A translation of some of these passages is to be found in Schuyler (1876), quoting Bretschneider (1875):

> In the next year (1219) a vast army was raised and set in motion toward the west. The way lay through the *Kin-shan* (Chinese Altai [Shan, or eastern end of the Tian Shan]). Even in the middle of summer, masses of ice and snow accumulate in these mountains. The army passing that road was obliged to cut its way through the ice. The pines and larch trees [probably spruce and fir] are so high that they seem to reach the heaven; the valleys (in the Altai) all abound in grass and flowers. The rivers west of the *Kin-shan* all run to the west, and finally discharge into a lake (Nor Zaisan [which may be modern Lake Balkhash]). South of the Kin-shan is *Bie-shi-ba* (*Bishbalik*, Urumtsi [Ürümqi]), a city of the *Hui-hu* (Mohammedans, Uigurs [the Uygur]). There is a tablet of the time of the T'ang dynasty, . . .
>
> At a distance of more than 1000 *li* [about 500 km, 330 miles], after having crossed the desert, one arrives at the city of *Bu-la* [not identified]. South of this city is the *Yin-shan* mountain, which extends from east to west 1000 *li*, and from north to south 200 *li*. On the top of the mountain is a lake (Sairam Nor [Sayram Hu]), which is over 70 or 80 *li* in circumference. The land south of the lake is overgrown with apple trees, which form such dense forests that the sunbeams cannot penetrate. After leaving the *Yin-shan* one arrives at the city of *A-li-ma* (Almalyk [Almaty]). The western people call an apple *a-li-ma* (*alma*), and as all the orchards around the city abound in apple trees, the city received this name. Eight or nine other cities and towns are subject to *A-li-ma*. In that country grapes and pears abound. The people cultivate the five kinds of grain as we do in China. West of *A-li-ma* there is a large river, which is called *I-lie* (Ili).

This passage is a little difficult to interpret, not least because of many changes in place-names. But it would seem that Genghis Khan and his army moved from what is now Xinjiang Uygur and through the modern city of Ürümqi (Bishbalik, which could be translated in Turkic as "five fishes," suggesting a place where fish from the streams was available—fish would have been a welcome change after the nutritional rigors of the Gobi). Interpreting further, the army took the Road of the North (Chapter 5), where, in a cold season, the ice and snow might have descended and barred the track in places even in spring or autumn. The city of Bu-la cannot now be identified, but the lake Sayram Hu lies about 100 km (60 miles) north of the modern city of Yining (old Kuldja) and just east of the Russia-China border. West of Almaty the army would have crossed not the Ili River, a formidable obstacle, but in fact tributaries of this huge river system, which all flow north and west into modern Lake Balkhash. There are now no groves of fruit forest south of Sayram Hu, only a vast region of intensive agriculture. The orchards around Almaty may have been grafted individual fruit trees but are more likely at this early date to have been selected, self-sown, open-pollinated seedlings, noted for their sweetness and fruit quality and thus transplanted and preserved. Similar seedling trees, on their own roots but with excellent and locally venerated fruit characteristics, can be found in that region today.

There is a much more recent glimpse of what may have been something close to the original, intact fruit forest. This picture comes from the revealing diary of Paul Nazaroff (1993), who was running from the Russian secret police, the dreaded Cheka (taking its name from a Russian acronym for "extraordinary measures committee"), just after the Marxist revolution of 1917. Nazaroff was a mining engineer and geologist who sought sanctuary in an area of Inner Asian fruit forest. Fortunately, he was very observant and of a broad and very well informed mind. Nazaroff's refuges are difficult to locate now but lay somewhere in the vicinity of Fergana, an administrative subdivision of what is now Uzbekistan to the southeast of Tashkent. He later described one of his hillside campsites as "a secluded spot in a valley of one of the tributaries of the River Choktal [Chatkal]":

> It is hard to say what was the origin of these fruit forests. Walnut trees are very widely distributed in the mountains of Turkestan, and undoubtedly are relics of the immense forests which covered the country and the Khirghiz steppe almost to the Urals in the Pliocene period. . . .
>
> It is not easy to account for the apples in these forests as within the memory of man there have been no gardens in the places where they grow now, and their fruit is in no way inferior in flavor to the cultivated sorts. . . . We can only conclude that it is a natural wild apple. The form of the fruit varies a lot: small and round, elongate, brown, yellow and red. Often the fruit is far too acrid to be edible, but sometimes sweet and juicy, even exceptionally so. . . . Some ripen at the end of June, others in July and August; and there are winter sorts too, which hang on the trees until the frosts. These are full of flavor.

From the early 1920s, the husband and wife team Mikhail Popov and Galina Popova, whose work was widely cited by Vavilov, collected in the Chimgan or Tchimgan region (Popova and Popov 1925, Popov 1929). Chimgan lies 80 km (50 miles) to the east of Tashkent and on the northern slopes of the Chatkalskiy Khrebet. They noted extensive tracts of wild apples and "that there are not two apple trees which are entirely similar to one another." It is interesting that they already referred to the sweet wild apple of the Tian Shan as *Pyrus malus* (that is, *Malus pumila*) and not *M. sieversii*. But the real extent of the destruction in the subsequent century can be gauged from Vavilov's comments in his paper of 1930:

> Not far from Tashkent entire woods composed of the wild apple may be observed . . . The wild apples found in the Caucasus are fairly small. The wild apples of Turkestan, however, and especially those of Semirechye, are characterized by their comparatively large size. The capital town is called Alma-Ata [the town of Verny], which means "City of Apples" since the whole town is surrounded by forests consisting of wild apple trees. When crossing the boundaries of Northern Tian-Shan, the traveler passes for a long way through woods of wild apple trees. Individual trees bear fruit which in quality is not inferior to that of cultivated forms. Some are of astonishingly large size and exceptional productivity.

Vavilov (1992) later wrote, "All around the city [of Almaty] one could see a vast expanse of wild apples covering the foothills which formed forests. In contrast to very small wild apples in the Caucasian [Caucasus] mountains, the Kazakh wild apples have very big fruits, and they do not vary from cultivated varieties. It was in 1929, the first of September, the time that the apples were almost ripe, one could see with one's own eyes that this beautiful site was the origin of the cultivated apple."

It is difficult now to reconstruct the distribution of prehistoric *Malus pumila;* virtually the entire area has been devastated by ethnic redistribution, collectivization, and wholesale environmental destruction through deforestation, agriculture, irrigation, the requirements of the military, the nuclear industry, and urbanization. But what does seem to emerge is that *M. pumila,* even before the sustained and devastating hand of humans, was sporadic in its distribution. The causes of this patchy distribution are not yet clear, but some suggestions can be made.

Today, wild *Malus pumila* is restricted to a few small intact forests of no more than several hundred hectares in extent in the Tian Shan, close to or on the border of Kazakhstan and China, in southern Kazakhstan in the mountainous

Khrebet Karatau, and with further small groves in southern Kyrgyzstan. West of Almaty there is little left of the fruit forest following the depredations of Stalin's and Khrushchev's ethnic cleansings and the so-called Virgin Lands campaign to convert large tracts into agriculture (C. C. Valikhanov in Akhmetov 1998). There are small intact areas, though, in the Khrebet Tarbagatay of southern Kazakhstan, lying directly over a major fault system (Maps 5 and 6; Hokanson et al. 1997, 1999). There are fragments of mixed fruit forest, with walnut (*Juglans regia*) and occasional *M. kirghisorum*, principally in Kyrgyzstan, and these relicts indicate the previous extent of the fruit forest and the diversity of the tree species associated with *M. pumila*. Ponomarenko (1990) reported that he found five new sites for what he called "*M. sieversii turkmenorum*" in the central part of the Kopet-Dag in 1983 (Map 8). These mountains stretch just west of Ashkhabad, Turkmenistan, on the border with Iran and close to the Caucasus Mountains. This possible area of distribution was noted by Vavilov (1930). The apple groves were found at a height of 1,600–2,200 m (5,250–7,200 feet), and it is suggested that further small stands might exist even farther southwest in the Zagros Mountains along the Iran-Iraq border. For obvious reasons in politically turbulent times, these areas remain to be extensively explored.

The fruit forest has been more thoroughly studied (Dzhangaliev et al. 2003) toward the eastern and northern ends of its current range, in Kazakhstan. There, the Dzungarian Alatau (Plate 17) and the Kopet-Dag are strongholds. These regions lie over or close to major fault lines, where the apple can sometimes be found with pomegranate (*Punica granatum*) and fig (*Ficus carica*). There were probably also considerable areas of fruit forest in what is now Xinjiang Uygur in China, in the Ili Valley and reaching down toward the city of Ürümqi, but little now remains there. In all areas, however, *Malus pumila* grows sporadically and in very different degrees of abundance.

In the warmest parts of the Pamirs of southern Tajikistan, *Malus pumila* can reach elevations as high as 3,040 m (9,970 feet; Ponomarenko 1990). However, farther north the lower temperatures reduce the altitude for potential growing areas, and in the Khrebet Zailiyskiy Alatau, south of Almaty, the upper limit may be as low as 1,900 m (6,200 feet). The common elevation range, with many small local variations, seems to be 900–1,600 m (2,950–5,250 feet). Above 2,700 m (8,860 feet), forests of spruce (*Picea schrenkiana*, Plate 15) and fir (*Abies semenovii*) are commonly found, and above that again to the tree line, before the snows are reached, thickets of several species of *Juniperus*. Only rarely is *M. pumila* found growing with the higher-level spruce and fir forests, and very rarely indeed close to the high *Juniperus* forests.

The Tian Shan and its outliers, that is, the Dzungarian Alatau (*alatau* meaning multicolored in Russian and possibly referring to the display of rocks of many types and thus different kinds of soil) and the Khrebet Tarbagatay, are not only very diverse in their geological construction and geomorphology but also in their soils. In the space of a short walk one can pass from deep, dark, rich, moisture-holding chernozem forest soils to sandy, dry, nutrient-deficient soils; yet all may be occupied by components of the fruit forest.

In the western Tian Shan, *Malus pumila* occurs mainly in the walnut forests of Kyrgyzstan, growing with *Juglans regia* up to 2,400 m (7,900 feet). It also over-laps with the distribution of *M. kirghisorum*. The apples form a secondary growth in these forests along with maples (*Acer turkestanicum* and *A. semenovii*), ashes (*Fraxinus raibocarpa* and *F. sogdiana*), hackberry (*Celtis caucasica*), pears (*Pyrus bretschneideri*, *P. korshimskii*, *P. turkomanica*, and *P. vavilovii*), several species of hawthorn (*Crataegus*), cherry plum (*Prunus divaricata*), but only occasionally the apricot (*Prunus armeniaca*). Sometimes there are also pistachio (*Pistacia vera*), almonds (*Prunus dulcis* and *P. ulmifolia*), buckthorn (*Hippophae rhamnoides*), and grape (*Vitis vinifera*). Here and there are small pure stands of a poplar (*Populus diversifolia*).

Sometimes the apple can be found at densities up to 80% in the fruit forest, but often it is accompanied by hawthorns (*Crataegus songorica* and *C. almaatensis*), apricot, and the maple *Acer semenovii* so that the apple forms a smaller percent-age. What is most striking is the diversity of the accompanying tree and large shrub species. In northeastern Kazakhstan and northwestern China (Xinjiang Uygur), *Malus pumila* is commonly found not only with the apricot but also wild pears (*Pyrus* spp.) and wild cherry (*Prunus* spp.). Wilfred Thesiger (1979) saw small and apparently natural forests of evergreen oak (*Quercus*), olive (*Olea euro-paea*), and walnut as far south as Nurestan, the mountainous province of eastern Afghanistan south of the Hindu Kush, and Badakhshan, in what is now extreme northeastern Afghanistan (Map 6), though he did not record a wild apple. There are suggestions here, too, of small pockets of relict forest, which again might be tiny remnants of the ancient widespread Tertiary forests.

In one small area, *Malus pumila* grows with *M. kirghisorum* but does not ap-parently often hybridize with it. There are suggestions that in a remote region, as far south as the region of the Wakhan Corridor of northeastern Afghanistan (Map 4), *M. pumila* grows with apricot and black mulberry (*Morus nigra*). The only common factor in this mixed fruit forest, which stretches east to west for more than 1,600 km (990 miles) and from north to south, in a broken pattern, perhaps 400 km (250 miles), seems to be *Malus pumila*. Such a distribution, with such diverse codominant species, suggests that what is seen now, at one brief point in evolutionary time, is the coalescence in a relatively benign period for

plant evolution of what were once discrete clusters of fruit forest, broken by earlier periods of climatic deterioration. The rapid mobility of the teardrop-shaped
apple seed, virtually undamaged in its transition through the jaws and gut of
horse, pig, or bear, may contribute to its universal and relatively rapid distribution. It would seem that the other species of shrubs and small trees have, with
their differing seed habits, mostly a slower rate of spread.

Today, near Urgut, Uzbekistan, only scattered trees can be seen in the Aman-
Kutan Nature Reserve, and extensive cattle grazing takes place at all elevations.
A few scattered apple trees can be seen on the northern side of the Fergana Valley, and these lie close to the edge of the coniferous forest zone. Bears are still
present in this region and may, in their autumn fruit feeding, assist seed dispersal. Notwithstanding the description by Paul Nazaroff of extensive fruit forest in
the early 20th century in the region of what must be the border area of what are
now Uzbekistan and Kyrgyzstan, very little in the way of continuous fruit forest
is now to be found there. Even within some of the special reserves, only isolated
mature trees, with little or no sign of natural regeneration, remain. But in degraded areas where cattle and sheep had been excluded, B.E.J. observed that
seedling apples erupted like weeds. The potential for reestablishing a seed-based
fruit forest always seems to be present.

An international team under L. J. M. van Soest (1998, pers. comm. 2002)
collected extensively in Uzbekistan in 1997. His impression and that of his collecting team was that although small, in-part artificially assembled orchards of
Malus pumila were not uncommon, nothing was to be seen that resembled natural fruit forest. It must be assumed that if a fruit forest still exists in Uzbekistan,
a region of Inner Asia where agricultural practices are the longest established, it
must be in some very small, isolated, north-facing valley, far beyond grazing and
gathering. Far to the east and close to Almaty, Kazakhstan, there are no groves
of wild apple trees to be seen at all, merely a very few scattered specimens on
isolated ridges. However, the wild apricot, which seems to be slightly more resistant to heavy grazing, is still fairly common there.

As explained in Chapter 1, apple seeds must not only be separated from the
placental tissue and exposed to cold-chill, sometimes as long as 200 days, for effective germination, they must also penetrate the sod or turf in some way. Survival of the seedlings depends on a number of factors, including the light regime
at the time of germination, growth conditions due to climate, nutrient availability, and avoidance of excessive browsing. Unlike seeds of the wild European apple
(*Malus sylvestris*), *M. pumila* seeds in the mountains of the Tian Shan do not usually fall on to established turf but are deposited either close to the mother tree, in
shaded but mixed forest vegetation, or on the edges of forest trails.

In the Tian Shan, B.E.J. saw considerable areas where cattle or sheep or goats were prevented from grazing and where apple seedling regeneration was prolific. In certain areas, however, it has been said that *Malus pumila* "propagates mainly by suckers up and down montane slopes. It often forms groves of root origin comprising up to 600 mature trees." (Alexander Sychov pers. comm. 1998). No such vegetatively proliferated groves have been recorded from the Dzungarian Alatau, nor northwestern Xinjiang Uygur, China, nor Kyrgyzstan, nor in the walnut forest northeast of Jalal-Abad. Nor have they been reported by any of the expeditions from Cornell University (see papers by Aldwinckle, Forsline, and Hokanson in the References). The general impression, in most areas seen by B.E.J., is that of seedling trees of varying ages and of great, almost infinite, diversity in growth form, posture, leaf texture, leaf angle, timing of leaf loss, pest resistance in a given season, bark texture, and so on (Plate 21). The conservation of this endangered yet economically important species should be of major international concern (Hokanson et al. 1999).

The odd distribution of *Malus pumila* has led to suspicion that a long period of human intervention has driven the development of these isolated areas of rich fruit forest. However, there is no evidence that there was any significant human activity in these high valleys and tumbling cliff slopes in the prehistoric or immediate historic period. Although archaeological and written records are almost nonexistent for this vast region, there would have been little or no incentive for nomadic groups, who depend on and are tied to large herds of grazing animals such as horses and sheep, which need constant protection, to venture into these high valleys. The only penetration would have been by occasional hunters in pursuit of boar, deer, or bear, and honey gatherers seeking the product of the rich bee populations. Using the yurt for habitation (Plate 22), such societies would have had little need for large timber from the spruce and fir of the high forest. The wood mainly used in the woolen felt yurt is willow (Plate 23), which could be gathered without difficulty from the lower reaches of the many streams and rivers flowing off the northern slopes of the Tian Shan. The only significant use of timber would have been for fuel. Apple wood burns well, but prescience would have dictated, at least in earlier times, that only apples of no culinary value, on the edge of the forest, be harvested for this purpose. This prudent behavior is no longer necessarily the norm (Plate 26).

Unlike that of the European apple, *Malus sylvestris*, only a little work has been done on population structure of the fruit forest. Few seedling trees in natural habitats have been studied for any length of time. What can be seen in the Tian Shan is that many specimens are massively multistemmed from the ground. At least in cultivation, some specimens may reach considerable size and age, though compared with many trees the apple is neither very large nor very long-lived. The largest that seems to have been observed in the Tian Shan is 16 m (50 feet) tall. In Virginia, however, a seedling apple is recorded as 21 m (70 feet) tall and 3.6 m (12 feet) in diameter (de Witt 2000). Most of the trees in the truly wild state in the Tian Shan seem to reach 5–8 m (16–26 feet) with a few specimens over 10 m (33 feet) and very occasionally an outstanding giant over 12 m (39 feet; Plate 20). Semi-dwarf specimens, particularly on poorer soils, are not uncommon.

Very many trees in cultivation in the West are recorded as having reached 100 years of age—the original 'Bramley's Seedling' just outside Nottingham, England, is about 200 years old but now produces only a small amount of fruit and is distinctly overmature. But the best-authenticated age record is that of an apple planted by Peter Stuyvesant in his Manhattan orchard in 1647; it was still healthy and producing fruit in 1866 when it was struck and destroyed by a de-railed train. Ages up to 300 years for seedling trees, both in the wild and cultivation under optimal conditions, seem not improbable.

In the Tian Shan it seems that in the natural state, following seed germination, initial vegetative growth normally lasts 6–8 years, followed by rising fruit productivity in the period to 10–12 years, with maximum seed production in the period to 25–30 years. Vigorous fruit production in what are believed to be natural conditions seems to continue in trees as old as at least 70 years. Subsequent fruit production and the age of the tree depends on the position of the tree in the stand (whether central or marginal) and the stability of the ground. Since rich stands of *Malus pumila* can commonly be found on extreme slopes, sometimes virtually cliffs (Plate 16), a lengthy occupation does not seem likely in what is a highly disturbed geological region. Such a relatively short life for any individual tree could give yet another stir to the evolutionary pot, contributing to the rich diversity of the apple. Could it be that the life expectancy of *M. pumila*, just a little over 100 years and in very rare cases up to about 300, is determined by the intense and long-established geological disturbance of the Tian Shan region? The longevity of, say, a bristlecone pine or giant redwood, measured in thousands of years, would be of little selective advantage in such an unstable region. A century plus, or thereabouts, may be an evolutionarily selected compromise between maximum seed productivity and probability of site retention.

Under the severe continental climatic conditions of the Tian Shan, seed production and germination, where grazing or browsing is limited, can be prodigious. Under controlled conditions, differences can be observed between the germination success of wild trees as opposed to selected cultivars. The best seed germination, 33.5%, came from wild seedlings, whereas seed from cultivated *Malus pumila* was 23.5% and that from *M. pumila* 'Niedzwetzkyana' only 20.2%.

In terms of numbers of apple trees per hectare, where seed germination and root-sucker production are fecund, initial frequencies may reach an astonishing 3,500–4,000 specimens per hectare (1,400–1,600 per acre). But after 2–3 years, 1,000–1,500 per hectare (400–600 per acre) remain, and this density falls to about 500–600 per hectare (200–240 per acre) as the fruit forest matures (Dzhangaliev 2003).

Notwithstanding the prodigious production of seed, in some areas vegetative propagation of *Malus pumila* is widespread. It is capable of vegetative renewal, spreading mainly by the formation of root suckers, to some extent by stem shoots, and also by adventitious root formation (Dzhangaliev 2003; Dzhangaliev et al. 2003). Such propagation does not develop exclusively as a result of damage to the trunk, shoots, or roots; whole, vigorous, undamaged trees have been seen to extend by vegetative means. Large populations of *M. pumila* occupy extreme slopes, and such slopes, with their constant threat of instability, may over evolutionary time have selected for this method of vegetative propagation. The constant exposure of new rock and soil surfaces may, like burning in certain Mediterranean and Australian forest habitats, clear the congested land for fresh apple seedings.

Root suckers can develop, and there can be root grafting between trees, but if the mother tree, with its shading crown, remains healthy and effective, the growth of the root suckers may be inhibited. But if the mother tree falls, some of the root suckers can reestablish the tree cover. Under somewhat rare conditions, horizontal, low apple tree branches may come to rest on the fresh soil litter and do what is known as layering propagation. At least two well-authenticated instances of self-layering of apples have been observed in Britain. Incidentally, this dispersal by vegetative methods is almost universal in the spreading, drooping growth of the black mulberry (*Morus nigra*), whose center of origin may not be far away from that of the apple.

The large size of some of the wild apple fruits of the fruit forest (Plate 21), compared to fruits of all other *Malus* species, has attracted skepticism. There are individual apple trees in the fruit forest with very large, sweet fruits that could be mistaken for cultivars of the orchard apple. Next to such trees, there may be ones with handsome but intensely sour, astringent fruits. The size of almost all the fruits of *M. pumila* in the fruit forest lies far outside the range of those of almost all other wild apple species, which tend to be 5–30 mm (⅕–1³⁄₁₆ inches) in diameter. Dramatic changes in size, alkaloid content, color, and sweetness have taken place in many crop plants within the historic period. Sometimes within only a few hundred years, not several million years, fruits, seeds, and tubers as diverse as the corncob (*Zea mays*), the potato (*Solanum tuberosum*), the tomato (*S.*

lycopersicum), and bananas (*Musa* spp.) have changed almost beyond recognition when compared with their wild ancestors. The cultivated apple is unusual in that its fruits are virtually identical to those of the ancestral plants.

A little of the genetic basis of the control of fruit size is now understood. A DNA region called a quantitative trait locus that controls fruit size has been located and cloned in the tomato (Frary et al. 2000). This quantitative trait locus has been transferred by genetic engineering back into ancestral forms and shown to engender dramatic size changes. Probably in only a few hundred years, the tomato fruit has been changed from one weighing only a few grams and less than 1 cm (about ½ inch) in diameter, probably distributed by small rodents and birds, to as heavy as 1 kg (2.2 pounds) and as much as 15 cm (6 inches) across. If such changes can take place in less than 6,000 years under the largely unconscious selection of human beings, it does not seem improbable that similar changes might have taken place in the millions of years available to the apples in the Tian Shan. Compare the fruits of *Malus baccata* (Plate 5), *M. hupehensis,* and *M. transitoria* with a modern *M. pumila,* for example, 'Discovery' (Plates 6 and 7). But what might these selection pressures have been and why do they seem to have been restricted to *M. pumila* in this particular region?

There is another feature of the fruit forest worthy of note. Reinforcing the diversity of fruit size and taste within the forest, B.E.J. frequently saw a diversity of tree form from round-headed to fastigiate, from dwarf through medium to substantial trees more than 10 m (33 feet) tall (Plate 20), and from unarmed trees to those with substantial thorns persisting into maturity. The existence of dwarf forms may have led to their selection later as rootstocks for use in grafting.

Notwithstanding the predilection of most browsing animals for sweeter, larger fruits, there are plenty of examples in the forest of bitter fruit that might, in later times, have served a cider industry (Chapter 7) but are difficult to explain in terms of sweet-toothed mammal selection. An interpretation of what at first seems to be a paradox may come from studies of fruit-eating birds, for example, bulbuls (*Pycnonotus barbatus*) and the blackcap (*Sylvia atricapilla*) in scrubland in Palestine (Izhaki and Safriel 1989). Ripe fruits of various species may contain tannins, compounds that often taste bitter, bind with proteins, and reduce the rate of assimilation of dietary nitrogen. Thus animals feeding exclusively on one source of tannin-rich fruit have an unsatisfactory diet and are forced to forage wider. A mixed pattern of apple fruits, with a range of tannin content from negligible to much, might thus tend to impose a much wider feeding pattern on their dispersers than would a forest of trees uniformly low in tannin and high in sugar.

The three phases of the apple crop

Another feature of the fruit forest may be interpreted from a study of apples in Western markets, showing that there are three consecutive phases in the kinds of apples produced through the growing season (Plate 6). Excluded from this speculation are the relatively modern triploids, which are generally completely sterile and thus could not play any evolutionary role in a natural fruit forest.

Phase 1 apples. As a general rule, early-season apples (July–August in the northern hemisphere) are brightly colored and often have a glistening skin, often predominantly red or with a red flush over a bright yellow ground, sometimes glaucous and frequently with a fine, particulate wax bloom. These wax blooms are found on the surfaces of many fruits at the point of perfect ripeness (Juniper 1995). They may contain flavonoids, which to certain animals and in certain light conditions, indicate ripeness, but the overall role of these blooms is not understood. Phase 1 apples are soft and thin-skinned, bruising very easily, so that most are not readily marketable. They are not very rich in flavor but are generally extremely sweet and juicy. The flesh is invariably soft and sometimes tinged or completely suffused with red. Above all, they are strikingly fragrant. An example from Russia is 'Red Astrachan', and British examples include 'Beauty of Bath', 'Discovery' (Plates 6 and 7), 'Gladstone', 'Irish Peach', 'James Grieve', 'Lady Sudeley', 'Laxton's Early Crimson', 'Stark's Earliest', and 'Worcester Pearmain'. Also classified here are the remarkable "transparent apples" (in German, *Klarapfels*), which have very fragrant and juicy transparent flesh. Among these are 'White Joaneting' (*joaneting*, "June eating"), known in England before 1600, and 'Grand Sultan' from Saint Petersburg, renamed 'Yellow Transparent' in the United States (Hanson 2005). True phase 1 apples do not keep well, and most are so soft they never come to market. 'Braeburn', 'Delicious', and 'Gala' come close to phase 1 apples in their characteristics.

Phase 2 apples. As the season progresses, the apples are less brightly colored, rarely if ever have a waxy bloom, and become harder-skinned, less fragrant, but richer, with a firmer flesh. They will generally keep for one to several months if correctly stored. Examples from British market shelves or orchards are 'Allington Pippin' (*pippin* from the French *pépin*, or seed), 'American Mother', 'Blenheim Orange', 'Cox's Orange Pippin', 'King of the Pippins' (Plate 6), 'Laxton's Superb',

'Lord Lambourne', 'Orléans Reinette', 'Ribston Pippin' (Plate 8), 'Sunset', and 'Winston' (Plate 9).

Phase 3 apples. From the end of the year come apples that are often very rich in flavor though lower in sugar, generally not fragrant, and with relatively dry pulp. Their skins are hard, dark, and frequently russetted (with minute corky, suberin protuberances), and the fruits often fall from the trees without damage. Often, they can be recovered from the leaf litter even after a British winter. These apples include 'Ashmead's Kernel', 'Belle de Boskoop', 'Brownlee's Russet' 'Claygate Pearmain', 'Cornish Gilliflower', 'D'Arcy Spice', 'Duke of Devonshire', 'Granny Smith', 'Hambledon Deux Ans', 'Ida Red', 'King's Acre Pippin', 'Leathercote' (Plate 6), 'Norfolk Beefin' (Plate 10), 'Rosemary Russet', and 'Sturmer Pippin' (Plate 11). Widely grown by discerning amateurs in western Europe, phase 3 apples are rarely seen in big markets because they lack the gloss seemingly preferred by the catering trade. Even after the much more severe continental winters of the high Tian Shan, fruits from the previous year can be found among the leaves on the ground (Dzhangaliev 2003). Phase 3 apples can recover from quite severe wounds, unlike almost all phase 1 apples, which immediately decay. They can hang on the tree late into the new year (Plate 35); they will generally keep for several to many months, often into the summer of the following year. Contrary to generally disseminated information in gardening magazines, many of these cultivars are cold resistant and will withstand temperatures down to −6°C (21°F) without damage.

If this pattern was duplicated in the natural fruit forest of Inner and Central Asia, as our observant voice from the past, Paul Nazaroff, indicates may be so, we might speculate about the significance of the three-phase interpretation as follows.

Early fruits are selected to attract potential and perhaps naive agents of dispersal. These vectors are principally bears. Bears, particularly when young, often climb into lower branches. There is therefore an evolutionarily selective advantage in attracting unaccustomed herbivores, such as the first-year young of bears. The attraction should operate over a short visual and olfactory range, hence the prevalence of bright, shiny colors, iridescent bloom, fragrance, and juiciness— liquid might be at a premium in the early part of the browsing year. It should be noted that all the proposed natural vectors of apple seeds belong to the so-called dichroics, animals with color vision restricted toward the red end of the spectrum (Regan et al. 2001). Some seemingly successful phase 1 apples are bright—shiny yellow or even white. True trichroics with full-color ability—human beings and some Old World apes—seem to have played no part in the early evolution of the apple. Therefore, the existence of the intense colors of the phase 1 apples remains a mystery.

As the season progresses and the vectors become accustomed to this food source, other factors might assume importance. Apples that fall from the trees extend the season of their availability. Their robustness and relative frost resistance ensure that they fall onto the now-accumulating leaf litter, generally without damage. They will then become available to bears just prior to hibernation, bears mature enough and perhaps sufficiently well nourished not to be over-inclined to climb trees. Apples will now be available to horses browsing on the edge of the forest and to other forest mammals that do not hibernate and cannot climb, such as boar and forest deer. Buried in the leaf litter and perhaps protected by an early covering of snow, the apple has its potential dispersal opportunities extended far into the new year by a wide range of ground-feeding vectors. Thus, through selection of tiny changes in its seasonal genetic diversity, the apple has its appeal, and thus success, maximized. However, if such seasonal tuning of the fruit to potential vector, an example of niche marketing in modern parlance, turns out to be correct, it could only have been be selected for in those plant communities where very high individual densities of a particular species were found. In certain areas of the Tian Shan, *Malus pumila* is thought to reach close to 80% of the canopy cover (Plate 17). In almost all other species of *Malus*, which are almost exclusively solitary, such a selection process would be irrelevant.

Far from the Tian Shan, managers of supermarkets are well aware of the attractiveness to the consumer of the brightly colored, sweet, glossy, fragrant, juice-rich, phase 1 offering, but the exigencies of long-range transport have ensured that modern raising of apples has combined most superficial phase 1 characteristics with the tougher skins and at least partial recovery from damage of the phase 2 and 3 types.

Do any other plants have mechanisms similar to that of *Malus pumila* whereby their distribution potential is maximized? Changes in pollination behavior are fairly common, as in the switch in the wild violet, *Viola riviniana*, from outbreeding (with chasmogamous flowers that open for pollination) in the early part of the season to inbreeding (with cleistogamous flowers that remain closed for self-pollination) late in the season if conventional cross-pollination has failed (Richards 1986). Are there any examples of fruit or seed change? Few species of seed plants reach densities comparable to those of *M. pumila* in the wild. The only example so far described appears to be the increase in caloric value, and the hardness of the seed coat, in a number of pines (Lanner in Schmidt and Holtmeier 1994, Lanner 1996). The caloric value of the pine kernels of both the whitebark pine (*Pinus albicaulis*) and limber pine (*P. flexilis*) markedly increase over the season (Mattson et al. 2000). A measurement of 4,800 calories was made in the seed of whitebark pine on 22 July, but this increased to 7,000 calories by mid-August. It is possible to speculate that the principal disperser of the seed, Clark's nutcracker (*Nucifraga columbiana*), is encouraged by the relatively high nutritional value but soft coat of the pine kernel and is sustained as a valuable disperser through the winter months. But the late-season seeds that are buried and forgotten, protected by their hard testas and well provided for by their dense endosperm, are the choicest of all.

Dramatic and rapid increase of fruit, seed, and vegetative size has accompanied the domestication of virtually every fruit crop in the post-Neolithic revolution. If *Malus pumila* did evolve entirely independently of humans, what special features of the Tian Shan, as opposed to all the other regions of the whole of the northern hemisphere in which other species of *Malus* are found, led to the development of the apple? What substituted for selection by humans?

Slowly and steadily, as the Tian Shan rose, its distinctiveness began to establish itself. The boundary of its vegetation was diffuse and changing in most places as the rainfall and heat patterns of each year or longer time period altered and the faults shifted. In the early preglacial period there would probably have been continuity of temperate forest east and west. Since then, an interface between foothill forest vegetation and steppe became established, the latter grading off into desert scrub and, finally, desert.

The Tian Shan is peculiar in one respect in that it represents a relatively benign (except in the long-term geological sense), isolated area surrounded by clearly defined, climatically hostile environments—deserts, essentially. The surrounding areas were rendered even more hostile by the onset of the last ice age about 1.75 million years ago (Maslin et al. 1998). Other regions of apple diversity—central and eastern North America, central and southeastern China, and central and western Europe—demonstrate no such clear boundaries. It is argued, in another context, that boundaries may be particularly important in the evolution of species (Schneider and Moritz 1999, Schneider et al. 1999). Based on evidence from mitochondrial sequences in a leaf-litter skink, it is hypothesized that evolutionary gradients, not isolation as is commonly supposed, may be a source of evolutionary novelty and perhaps new species. Over the entire period of the Quaternary (Table 1), the whole region of the Tian Shan would have been rich in changing gradients and boundaries of every nature.

Relatively recent research may provide both a vector and some rough estimate of the time of the redistribution of apple seed of whatever species. There exists, in a very disjunct pattern, a member of the crow family, the azure-winged

magpie (*Cyanopica cyanus*, Plate 13). It is found in southwestern Europe, principally along the coast of Portugal and western Spain, where it can often be seen
feeding in the canopies of the pine forests. And in Asia, on the southern islands
of Japan, in Korea, and in eastern and southern China reaching the Gansu Corridor, the magpie is common in small-fruit trees. The two widely separated populations appear, in every other respect, to be identical.

It had previously been supposed that the azure-winged magpie is truly an
East Asian species that was introduced and naturalized in Iberia at an early and
unknown date, perhaps by some Portuguese ship's captain. From the 16th century, the Portuguese were certainly trading in Asia into the edges of this magpie's
distribution. However, fossil bone discoveries in Gibraltar indicate that the magpie was certainly present there in the Pleistocene. Fossil bones have also been
found in China, indicating that the species was a resident there over a long period
of time (Cooper and Voous 1999, Cooper 2000).

Evidence has since emerged that actually link the two populations in ancient
time. Studies of the evolution of the sequences of mitochondrial cytochrome b
and ND5 genes of both bird populations suggest that they were in contact but
may have diverged as early as 3 million years ago (Cardia et al. 2002, Fok et al.
2002). If these two populations were separated in the late Pliocene, it can be
speculated that, at least in the West, the advance of the last glaciation broke the
continuity with Asia. Iberia remained a refuge for animals and plants through
each successive glaciation (Willis 1996). Subsequently, the Tian Shan itself became isolated, not by glacial ice, which did not approach these ranges (Map 1),
but by progressive desiccation, which continues, resulting in the formation of the
Gobi and Taklamakan and the constant rain of loess from the west, over much
of central and eastern China.

In the autumn fruiting months, the azure-winged magpie of Asia can be
seen chattering and feeding voraciously on small-fruited *Malus* species, including
M. baccata (and *M. mandshurica*), *M. hupehensis*, *M. transitoria*, *M. yunnanensis*,
and larger-fruited modern cultivars, including those of *M. pumila* and species
hybrids in the national collections and botanical gardens of eastern China. The
bird does not now, if it ever did so in historical or early prehistoric time, occur in
the Tian Shan (D. Taylor 1999), although the related black-billed magpie (*Pica
pica*) is common. However, it is possible to speculate that ancestors of the azure-
winged magpie, and perhaps similar fruit-eating birds, occupied the forest corridor that extended from western Europe to eastern Asia in the late Tertiary and
early Quaternary. Such a corridor would be consistent, too, with the Beringian
continuity into North America. In the west, the continuous corridor through
which this magpie moved was eventually broken by the advancing ice, in the east

by progressive desiccation, and in the Tian Shan itself by the lifting of the mountains above the natural range of the magpie or its ancestral forms. But *M. pumila* in the Tian Shan today, in contrast, is almost entirely distributed by mammals, principally bears and horses.

Dispersal of apple seed by bears and horses is well documented. Northwestern Europe in the immediate postglacial might not have possessed the steppe and grassland sufficient to lure back the wild horse, but could some species of apples, at least in northwestern Europe and possibly elsewhere, have, initially at least, been distributed by wild cattle? Cattle might have been better suited to the mixed woodland, moraines, and damp postglacial floodplains of the region. Is there is evidence that such cattle existed in early postglacial Britain and northern Europe? Do such have any relevance to the distribution and evolution of *Malus pumila*?

A remnant, some 35 in number, of ancient, large, white cattle still survive in Britain and are known as the Chillingham herd or White Park cattle. This small herd has been at Chillingham near Alnwick in Northumberland in northern England, probably for the last 700 years. They were almost certainly enclosed by the owners of Chillingham Castle in the 12th century, and they are now the property of the Chillingham Wild Cattle Association. Small subpopulations of the herd now live on Queen Elizabeth II's land near the Moray Firth in northeastern Scotland, and another herd has been established in the United States. A number of features suggest that though they are closely related, they are not truly direct descendants of the ancient wild cattle, the aurochs, that roamed Europe up until the advent of the Neolithic revolution in agriculture some 10,000 years ago. The Chillingham herd is a remnant of escapes of very early domesticated cattle; examples are known to roam other parts of Europe. Nevertheless, there is archaeological evidence that at least small herds of the closely related ancient wild cattle were still present in Britain at the beginning of the construction of Stonehenge in Wiltshire some 4,300 years ago.

The word *cow* conjures up an image of a placid milk machine in a meadow, but nothing could be further from the truth so far as the Chillingham herd is concerned. These animals are noted for their ferocity and have been responsible for several well-documented human deaths, including, very nearly, King Richard I of England. They possess long, lyre-shaped horns, which they have no hesitation in using. They demonstrate, in their collaborative defense, particularly of their young, that they would have had no difficulty protecting themselves against the modest carnivores—wolf, fox, wildcat, perhaps even the occasional bear—of prehuman, Paleolithic, Neolithic, and Bronze Age Britain, and western, central, and eastern Europe.

During the interglacials and immediate postglacial, the wild cow, almost certainly with the wild bison, must have ranged throughout the whole of Europe to the boundaries with Asia. But there is no evidence that either was ever present in the Tian Shan, and since both chew the cud, neither would have been major distributors of defecated plant seed, though with their hooves they both would have forced rejected fruit seeds deep into the soil.

Enter the bear

As the Tian Shan began to rise, the nature of its geology would have provided an Elysium for the bear (*Ursus arctos*) and certain other forest-dwelling animals. The unglaciated mountain slopes, below the snowpack, are a perfect habitat, particularly where caves are washed out of limestone rock (Plate 16). In contrast, the gentle, thick-loess-coated hills and low mountains of western, central, and southeastern China or the flat, high Qinghai-Tibetan and Mongolian Plateaux present no such rich possibilities. Deep, soft, loess soils may be good for certain agricultural techniques but provide neither much in the way of secure, stable cave dwellings for winter hibernation nor the immense diversity of habitat on which omnivorous animals depend. Likewise, a depressingly uniform habitat, both topologically and in terms of soils, however rich in an agricultural sense, provides little stimulus for the sustained evolution of a species. At least orogenically, life in the Tian Shan was never dull.

Research indicates a rich diversity in the diet of Tertiary bear species. Their claws evolved, enabling them not only to catch fish and tear open bees' nests but also serving as excellent rakes for the harvesting of fruits of all kinds. Thus, according to an analysis of the isotopes of their fossil bone collagen, Pleistocene short-faced bears (*Arctodus sinus*) of eastern Beringia were very carnivorous (Matheus 1995). However, modern bears of the genus *Ursus*, including grizzlies (*U. arctos horribilis*, which coexisted in some areas with *A. sinus*), are distributed in a complex pattern of species and local variants all over the northern hemisphere. With *U. malayanus* they penetrate into subtropical and tropical regions as far south as the equator. Without exception they seem to

The Tian Shan bear.
Drawing by Rosemary Wise.

be comprehensively omnivorous. These bears include, at least in their spring and autumn diets, a wide range of plant material, among which are mosses, horsetails (*Equisetum*), and roots, corms, leaves, nuts, and seeds, including so-called pine nuts, but also almost invariably berries and other fruits of many species. In spring, the bears of the Tian Shan can be seen feeding on wild rhubarb and juniper berries (Anna Pavord pers. comm. 2003). Brown bears of almost every area, if sources are available, scavenge extensively on bee larvae, combs, and honey. A. A. Milne, author of *Winnie the Pooh*, was right.

North American brown bears (*Ursus americanus*) and grizzlies, which consume wild fruits for energy accumulation in the late autumn prior to hibernation, are constrained by several factors, including intake rate, the physiological capacity of the gastrointestinal tract, and the metabolic efficiency of the gain in body mass (Welch et al. 1997). Thus, whether in captivity or free-ranging, it seems that bears depend on large fruits or large fruit clusters to gain weight rapidly. Interpreting these findings, there appears to be positive selection for seeking out, feeding from, and remembering the sites of fruit-bearing trees that have high concentrations of large, attractive fruits. If the speculations concerning the three phases of apple fruits are correct, positive and reciprocal selection for the nutritionally richer phase 3 fruits would also be advantageous.

The brown bear (*Ursus arctos*) found in the Tian Shan is legally protected over almost all its range. Nevertheless, it remains a target for trophy hunters (Plate 24). The feeding range of the bears in the Tian Shan is not known, but grizzly females in the conifer forests of mountainous western North America are known to range over 600–1,000 km^2 (230–390 square miles; Mattson et al. 2000). What is more, unlike the grizzly, brown bears climb trees. Usually, as is commonly observed in the Sierra Nevada of California, the younger and more agile members of the brown bear group are persuaded by their mothers to climb the trees in search of suspended backpacks. A delightful illustration of this division of labor is illustrated by an unknown artist, about 1570, in a depiction that is part of the *Romance of Amir Hamza*, painted for Akbar, the Mughal emperor of India. A mother honey bear (*U. malayanus*) is shown holding fruit while a younger member of the family climbs the tree (Lucie-Smith 2001).

Bears learn quickly. The brown bears of Montana are known to seek out and feed on the sweet fruits of grafted apple trees (imported *Malus pumila*) in orchards and spread the seed around in their feces, "where hungry bears play Johnny Appleseed" (McGahan 2001). In fact, as explained in Chapter 7, it is the converse: Johnny Appleseed was behaving like a bear.

Seeds of many species of plants pass relatively uninjured through the guts of many mammals. Of 27 fruit-bearing plant species passed through animal guts in the Mediterranean, fewer than 1% of the defecated seeds could be seen to be damaged (Herrera 1989). Evidence for the transit of 73 species of plants through the guts of 28 species of nonflying mammals representing 18 different families has been reviewed (Traveset and Wilson 1997, Traveset 1998) as has gut transmission of seeds not connected with fruit eating (Pakeman et al. 2002). In summary, the majority of seeds in transit showed either a neutral or enhanced level of germination.

Transit time through the gut is usually about 48 hours but varies greatly. It can be as long as several days, and occasionally horse transit is measured in months (Janzen 1982). From several hours to a day seems to be the measure in the brown bear (Traveset and Wilson 1997, Traveset 1998). However, it is presumably also possible that a small number of seeds might remain in the gut of a hibernating bear, since hibernating bears do not defecate.

As a generalization, small seeds pass through the gut more rapidly. Thus one could speculate that thick, fat apple seeds, which are commonly more viable, would travel the farthest from the mother tree. Small apples can be seen in bear feces, having passed intact through jaws and gut (Herb Aldwinckle pers. comm. 2001). The seeds therein will almost certainly not germinate. Seeds from the larger apples, however, are separated by the animals' jaws from the placental tissue and deposited over vast areas of the Tian Shan (Plate 25). Thus there is, and presumably has been for millions of years, positive selection for larger, sweeter fruits of *Malus pumila*. However, the apple seeds in the Tian Shan are smeared over the surface of the forest soils. In the mountains there are no large ungulates with sharp hooves, such as cows, to force the seeds through a grass mat into the damp soil below. The wild boar, with its all-year-round, well-attested, soil-churning, root-, tuber-, and bulb-seeking habits, may assist in burial. Whatever the mechanism, as can be seen in places where areas are protected from forest-floor grazing, seed germination can be prodigious.

Enter the horse

The horse family evolved in North America and only entered the Old World some 2 million years ago during one of the periods when the Bering Strait land bridge was in existence. The advance of the glaciers in the West not only obliterated vegetation but also opened the Beringian land bridge as water was locked up in the ice sheets. The concomitant spread of the deserts would also cause the contraction of forests but extend the grassland and steppe on which wild horses could range.

Equines spread across Asia and on into Africa, separating in evolutionary time into the horses on the steppes north of the Tian Shan, which were probably the ancestors of all modern horses (*Equus caballus,* Plate 28), and the half-asses or onagers (*E. hemionus*) in Mesopotamia and Iran, the asses (donkeys) in Palestine and northeastern Africa, and the extinct quaggas and extant zebras in southern Africa (Vila et al. 2001). The true wild horse, or its near relatives, the Tarpan and Przewalski horses, may still survive in the remoter parts of Kazakhstan (Bökönyi 1974).

Strangely, horses became extinct in North America perhaps some 8,000 years ago (Clutton-Brock 1981). They may have been wiped out by the so-called Clovis people, the first human invaders of the North American continent. It would have been extremely difficult for the peoples of North America to have domesticated the horse there even had they so wished. There were neither wild nutritionally rich grains such as oats, nor peas or beans, in the northern half of the continent that could have sustained domesticated, range-restricted horses in working condition through a continental winter. Only the complete freedom of long-range grazing and browsing in all seasons made their existence possible. The horse, in a domesticated form, was not introduced into North America until the Spanish colonial period, beginning in the 16th century. The so-called wild mustangs of the American prairies are descended from relatively recent escapes, probably from early Spanish settlers on the West Coast (Sherratt 1984, 2004).

In Asia, horses or their predecessors would have penetrated right through to the boundaries of western Europe. Fossil remains are widespread. Bones of wild horses, dating from about 30,000 or possibly as long as 50,000 years ago, have been found among the more spectacular remains of woolly rhinos in a sand and gravel pit at Whitemore Haye in Staffordshire, England. There is plenty of evidence that in the Paleolithic, horses were a common source of food for humans (Sherratt 1984, 2004). But these temporary interglacial incursions of the horse into Europe would have been halted by the periodically encroaching ice sheets that scraped northern Europe more or less clear of vegetation and almost all other living things, at least four times in the recent geological period. That the wild horse did repopulate western Europe at certain times in the Quaternary is also attested by the beautiful depictions on the walls of the cave of Chauvet-Pont-d'Arc in southeastern France, drawn 30,000–33,000 years ago. Others, about 15,000 years old, are preserved in the Lascaux caves in southwestern France. Horses would have been driven back into their Asian steppe strongholds from these areas from time to time as successive sheets of ice advanced.

In their first invasion of Asia, horses spread south and west, and passed along the edges of the temperate forest corridor. They may have assisted in the movement of the early apple species out of central and southern China into the rising mountain slopes of the Tian Shan. Inevitably, they would have stirred the gene pool of early apple species.

The wild horse of the steppe, evolving teeth to feed on the grassland and with the speed and maneuverability required to survive in that habitat, would not have willingly browsed into the deeper woodland—dangerous predators under the closed canopies and in the tangled shrub vegetation would have kept it wary. Nevertheless, as the quality and quantity of the plains vegetation, principally species of grass and a small number of otherwise edible herbs, weakened toward the end of dry, hot, continental summers, the horses, desperate for sustenance and moisture, must have penetrated into the woodland edge. Such a browsing pattern would have coincided with the ripening of the huge range of fruits— apple, pear, plum, apricot, and hawthorn—of the fruit forest. The horses would have fattened on the rich fruit at the edge of the forest that, in common with the autumn diet of bears, would help sustain them through the severe winter. Unlike a bear, however, hibernation was not an option for a horse.

The horse's speed and range, combined with a long gut-transit time, may have been significant in spreading fruit seed, particularly apple seed, from one area of fruit forest to another. As suggested in hypothesis 3, horses may have provided tenuous but significant links between portions of the fruit forest as it was fragmented by climatic or orogenic changes, or both.

It seems possible, too, that the horse's well-known appetite for apples, combined with contraction of the fruit forest, might account for isolated patches of small sweet apples in the Caucasus Mountains, the Crimean Peninsula, parts of Afghanistan, Iran, and Turkey, and the Kursk region of European Russia. These populations, dating from perhaps as long as 2 million years ago, might have been spread by the horse.

Under the forest canopy, other mammals would have enjoyed the products of the fruit forest. The fate of vast quantities of apple seed, and inevitably many other seeds, as discussed before, is to pass through an animal's gut. Most frequently, it is that of the bear, but it can be that of the horse in early summer, and other animals have a role. Such a transition will almost certainly enhance germination and, just as importantly, carry the seed burden well away from the mother tree. Small, agile deer and the wild boar, particularly in winter, would have played a part in spreading and, in the latter case, possibly inadvertently, burying apple seed. Along the river valleys, certainly in region of the Ili, the Bukhara red deer (*Cervus elaphas bactrianus*), the omnivorous honey badger (*Mellivora capensis*), and the roe deer (*Capreolus capreolus*) would have inhabited at least the edge of the forest. At home under the canopy would have been musk deer (*Moschus* spp.) and muntjac (*Muntiacus* spp.), still present in the remoter areas of the Tian Shan. All these would have eaten fruit. For some, notably the honey badger, hibernation in the depth of winter was an option. Most of the other mammal inhabitants were adapted to eat and drink through winter.

Camels

The center of evolution and the place of the domestication of the camel are disputed. It is believed that very small herds of the wild two-humped camel (*Camelus ferus*), from which the domesticated Bactrian camel (*C. bactrianus,* Plate 28) is thought to be derived, still survive in the remoter regions of the Gobi. Possibly there were two centers of evolution, namely, northeastern Africa and central Asia. It seems possible, too, that the single-humped camel evolved from the two-humped camel as heat tolerance was selected for (Bulliet 1975, Potts 2004). Hybrids between the two are common.

However, the efficacy of the camel in the distribution of the apple or any other seed is dubious. The camel masticates so fiercely and regurgitates so regularly that every scrap of vegetable matter is reduced to pulp; every pip or grain is so degraded that germination of even the smallest seed after passage through this gut becomes highly improbable. Camel droppings are so dry that they are combustible as soon as voided, and throughout history they have enjoyed a premium as fuel for fires. Camel droppings, too, are so round and hard, a feature of the spoor of many desert-dwelling animals, that they are sometimes used instead of pebbles as counters for ground games like the European Nine Men's Morris (Lewin 1999). Neither of these secondary uses is conducive to the spread of propagules of any kind except those of viruses, bacteria, and fungi.

And dung beetles?

The possible role of dung beetles in the distribution and burial of seeds needs to be examined. Nothing seems to be known of the dung beetles in the Tian Shan, but since they exist on every continent except Antarctica and are indigenous even in New Zealand, it must be expected that they are there, too. There are more than 7,000 species worldwide. Under tropical conditions their achievements can be prodigious. A 1.5-kg (3.3-pound) pile of elephant dung on the African savanna attracted 16,000 dung beetles of different species that ate or buried the dung completely in 2 hours. Every sort of dung is buried somewhere, and humans

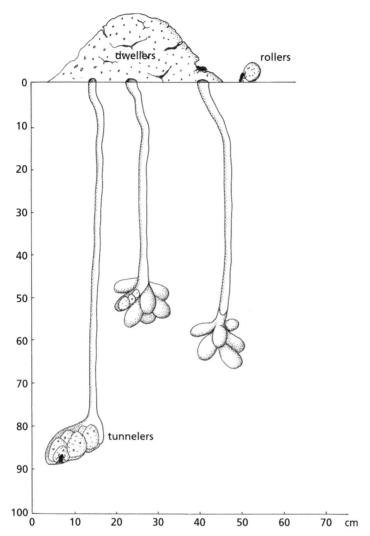

Dung beetles: dwellers (top), rollers (top right), and tunnelers (bottom).
Drawing by Rosemary Wise, based in part on Hanski and Cambefort (1991).

should be grateful for this frenetic activity. Dung beetles can conveniently be
divided into rollers, tunnelers, and dwellers; it can safely be assumed that all
three are present in the Tian Shan. Rollers ball up the feces, often into spheres
far larger than the beetles themselves, roll them away, dig tunnels, lay one or
more eggs within the dung ball in the tunnel, and then backfill it. Tunnelers dig
underneath the fecal pile and often build a network of tunnels, dragging pockets
of dung into the side chambers wherein they lay their eggs. Dwellers live entirely
within the fecal mass, and some feed not upon the residual vegetable and animal

matter but simply upon the liquid component. Rollers have been known to move a dung ball 400 m (1,300 feet), and tunnelers to bury dung balls with their eggs as deep as 1 m (3⅓ feet; Hanski and Cambefort 1991). However, it must be emphasized that nothing is known of the species nor the activities of dung beetles specializing in bear, horse, or wild pig dung in the Tian Shan.

Very little seems to be known of the fate of fruit seeds in general as a result of this dung beetle recycling, although some beetles in other habitats separate out the larger seeds and remove larger pieces of vegetation before depositing their eggs. But it seems plausible that the majority of robust apple seeds would survive, and thus particularly as a result of the activity of rollers and tunnelers be placed in conditions potentially advantageous for germination. Moreover, their activities would tend to move the fecal and seed mass away from the parent tree, thus reducing competition.

PLATE 1. *Malus pumila* from the collection *Libri Picturati,* dating from about the middle to the second half of the 16th century. These watercolors, of which about 1,860 survive, are bound into 13 white vellum volumes. They are of garden plants, crops, wildflowers, conifers, mosses, liverworts, and lichens, including some plants from the New World. The artists, of whom there are certainly several, are unknown, but the illustrations may have been prepared in what is now Belgium. Their precise history remains mysterious (Zemanek and de Koning 1998). Reproduced by kind permission of the Director of the Jagiellonian Library, Krakow, Poland.

PLATE 2. *Malus pumila* in flower. Watercolor by Rosemary Wise.

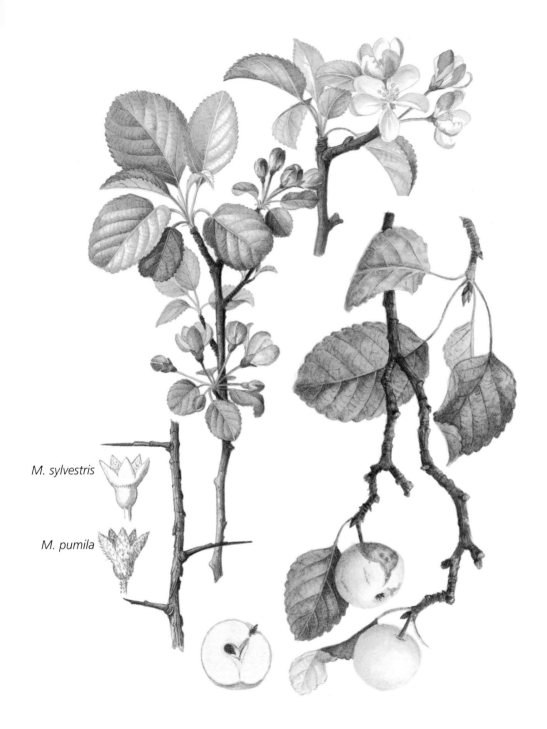

M. sylvestris

M. pumila

PLATE 3. *Malus sylvestris* and *M. pumila,* with calyx inserts to show those differences in their flowers. Watercolor by Rosemary Wise.

PLATE 4. *Malus kirghisorum.* Watercolor by Rosemary Wise.

PLATE 5. *Malus baccata,* a bird-dispersed apple. Watercolor by Rosemary Wise.

PLATE 6. The three phases of the apple crop: 1 (upper, *Malus pumila* 'Discovery'), 2 (middle, 'King of the Pippins'), 3 (lower, 'Leathercote'). Watercolors by Rosemary Wise.

PLATE 7. *Malus pumila* 'Discovery', a phase 1 apple. Watercolor by Rosemary Wise.

PLATE 9. *Malus pumila* 'Winston', a phase 2 apple. Watercolor by Rosemary Wise.

PLATE 8. *Malus pumila* 'Ribston Pippin', a phase 2 apple. Although a triploid, its fruit is no more than medium-sized. Watercolor by Rosanne Sanders.

PLATE 10. *Malus pumila* 'Norfolk Beefin',
a phase 3 apple. Watercolor by Rosemary Wise.

PLATE 11. *Malus pumila* 'Sturmer Pippin',
a phase 3 apple. Watercolor by Rosemary Wise.

PLATE 12. *Malus pumila* 'Niedzwetzkyana'.
Watercolor by Rosanne Sanders.

PLATE 13. The azure-winged magpie, *Cyanopica cyanus*. Watercolor by Rosemary Wise.

CHAPTER 3

Archaeology and the apple

THE MIGRATION OF HUMAN BEINGS across Asia involved penetration deep into China from the West, but the peoples now living there arrived too late to effect the dispersal of the apple westward. Wild horses, dispersers of apples, were domesticated about 6,000 years ago, but the first possible record of the sweet apple in the Middle East is only 3,800 years old, and unambiguous records date only from the time of Alexander the Great. In Classical texts, notably the Bible, the apple has been much confused with other fruits.

WHETHER OR NOT the movements of early people to and fro across Beringia (Map 1) had any influence on the movement of plants and animals to and from North America is debatable, but certain facts are worth recording. From the evidence of human mitochondrial sequencing, there appear to have been at least four human invasions of North America from eastern Asia. The source of one can be traced to Siberia and northeastern Asia, notably in the vicinity of Lake Baikal and the region of the Altai Shan of western Mongolia to Kazakhstan and the Sayan Mountains of southern Russia to the north. Another source is the Jomon or prehistoric peoples of Japan, which may have split to give both the present-day Inuit and, on the other hand, the Blackfoot and Iroquois tribes. And another strand can be traced back to what is now China (Greenberg and Ruhlen 1992). Any or all may have contributed to the spread of the genus *Malus* in North America. The use of wild *Malus* species by the indigenous American Indian population is not documented as widespread or extensive, though Moerman (1998) has listed recent uses.

An understanding of the pre-Columbian human population of North America is complicated by the fact that there are signs, indicated by presence of the so-called mitochondrial haplogroup X, that European populations similar to groups now found in the Middle East migrated to North America about 28,000–30,000 years ago (Lewin 1998). There is speculation that some of these movements may have come from Australasian regions (Greenberg and Ruhlen 1992, Hey 2005). The pattern is further complicated by the discovery in southern Washington State of what appears to be a Caucasoid individual, the so-called Kennewick Man, injured by a Stone Age spear or arrow point and whose remains are about 9,300 years old (Huckleberry et al. 2003). DNA analysis of his possible affinities to other North American groups or even Europeans has not so far been possible. Tentative investigations put his relationships among the southern Indian peoples or possibly the Ainu of Japan, Sakhalin, and the Kuril Islands. These groups, though, seem unlikely to have effected the spread of what seems to be an Asian-derived *Malus* group. There does not seem to be any evidence of a backflow of *Malus* genome to Inner or Central Asia; the Pyreae of North America appear to be quite distinct. At some point, however, very possibly before the glaciations, the horse seems to have migrated from North America to Asia by this route, whether assisted by humans or independently.

There are scant but detectable records in the Tian Shan area of Paleolithic humans. Early humans, perhaps pressed by increasing population and the finite number of caves in the Tian Shan, moved out into the foothills and plains. Nevertheless, the nature of the Tian Shan, with its dramatic geological history, must have provided Paleolithic humans with a wide range of potential dwellings. Such an area would have contrasted with the loess regions to the east, where the soft, unconsolidated soils offered no such convenient shelter. So it is likely that early humans would have been familiar with the Tian Shan, at least in the summer and autumn. In this region, too, and possibly with the onset of the domestication of the horse and the camel, it can be speculated that fishing, hunting, and gathering humans moved out of the caves and invented the far more effective and comfortable mobile cave, the yurt (Plates 22 and 23).

But long before recorded history, sophisticated nomads from the west, probably Indo-Europeans from the Caucasus Mountains, penetrated far eastward into these remote parts of Asia (Stein 1903, 1907, 1921; also see Allen 1996, Walker 1998, Barber 1999, Mallory and Mair 2000). Deep in what is now the Taklamakan, Sven Hedin, the late 19th and early 20th century Swedish explorer, found lost towns, including Karadong (Map 6). These towns had been overwhelmed by sand but could still be detected by the remains of avenues of poplar trees, with orchards of plums and apricots. Mark Aurel Stein, digging at Niya, also found avenues of white poplars, and orchards of apple, plum, peach, apricot, and mulberry, the wood of which his laborers recognized from their own villages far away on the fertile edge of the desert. It is impossible now to say whether these were simply transplanted seedlings or grafted specimens.

Stein identified modeled figures of Pallas Athena bearing an aegis and thunderbolt, portrait heads of men and women with barbarian (not Han Chinese) features, a standing and a seated Eros, and Heracles and other Athena figures. Many of these artifacts strongly suggest contacts with Greek civilizations at an early date. Other artifacts, such as images of elephants (Allen 1996), give proof of contact with Indian civilizations. Skeletal remains indicate that horses were in use in this society, even on the edge of the desert.

So far, such settlements have only been found around the fringes of today's deserts, where an extraordinary combination of circumstances has preserved human remains with artifacts. The bodies are not mummies in the generally accepted sense, since they were not prepared in any specific way for preservation. But the combined effects of winter cold, desiccation, and salinity have preserved objects of extreme delicacy, such as fabrics, to an extent unmatched almost anywhere else in the world. These sophisticated Bronze Age to early Iron Age peoples wove advanced textiles, including twills, and could tan a range of leathers. As attested by the body of a little boy in the history museum in the Turpan depression of Xinjiang Uygur (Map 4), they even knew how to fashion an artificial teat as a surrogate for breast feeding (Mallory and Mair 2000). Some artifacts are 4,000 years old, but some may be far older.

These people have been named the Tokharian, or Tocharian, but the name is somewhat misleading as it implies some distinctly separate origin when it is probably best to think of them as a branch of the widespread Caucasians (Indo-Europeans), speaking an unknown language (C. Renfrew 1989). Notwithstanding their ethnic and linguistic origins, they provide a dramatic example of the movement of sophisticated agricultural peoples, growing fruits with other crops at least 4,000 years ago, deep in what is now China and, as evidenced by their artifacts, obviously in contact with Europe and distant parts of Asia (Map 7).

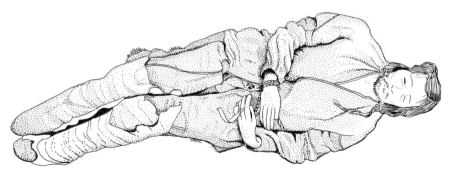

Mummified body of a Tokharian man about 50 years of age at death, estimated to be 4,000 years old, preserved in the museum near Ürümqi, Xinjiang Uygur. Drawing by Rosemary Wise from a photograph by Jeffery Newbury.

The Turkic Corridor

The ebb and flow of humans across the great expanses of Inner and Central Asia defy easy summary. But in the prehistoric period, Indo-Europeans traveled from the west deep into what is now China. However, there is an apparent mismatch between the peoples who now occupy the areas of the east–west land routes and the supposed roles of other ethnic groups in the apple's journeying. Today, what is called here the western migration route of the apple (and certain other fruits) is more or less completely occupied by Turkic-speaking peoples. This is the corridor of the "turbaned peoples" (Plate 29), extending from the Gansu region of China and from the lake Qinghai, formerly called Koko Nor, in the east (Map 6; Fleming 1934), west to the Mediterranean, with outliers in northern Siberia. The Hui, Kazakhs, Kyrgyz, Tatars, Turkmen, Yakuts, Uygur, Uzbeks, and Turks speak Turkic languages. Zahīr-ud-Dīn Muhammad, known as Bābur, whose military conquests, beginning in 1497 in Samarqand, established the Mughal empire, spoke a dialect of Turkic. This Turkic-speaking corridor was established, at a very rough estimate, no more than 1,000 years ago. Thus the presence of these Turkic-speaking peoples is too late to have influenced the initial migrations of the apple and related Tian Shan fruits, according to our hypotheses, which concentrate on the period from about 7,000 years to 2,000 ago. Nor were the Turkic peoples connected with that seminal discovery, grafting (Chapter 4), which was centered somewhere in valley of the Tigris and Euphrates about 3,800 years ago.

Nonetheless, the efficacy of the Turkic Corridor as a linguistic corridor can be seen by examining the modern Turkish word *turfanda*. In the markets of Turkey the word is used for expensive, imported, early-in-the-season fruits and vegetables. A probable interpretation is that at least some of these delicacies came from the area of the Turpan, or Turfan, depression, where ancient underground water channels allow early ripening of cultivated fruits (Chapter 5).

Moving a word is one thing, but moving a product or technique of advanced agriculture is quite another. On the whole, new ideas, new crops, and new patterns of cultivation move relatively slowly through settled, conservative, agricultural communities. Could what we observe along the Turkic Corridor now reflect the imposition of agriculture on what were in prehistoric times exclusively migrant and nomadic peoples? Such enforced settlement has occurred as recently as the 20th century, notably under Stalin and markedly after the German invasion (C. C. Valikhanov in Akhmetov 1998). Prior to the regimes of modern despots, this great corridor, which probably represents what in prehuman time were the migration tracks of grazing animals, was the realm of nomadic peoples who had highly developed skills of movement, transportable housing, and a horse culture. It was along this same corridor that the apple and other fruits moved westward.

An apple seed in the gut of a horse might move 65 km (40 miles) in a single day. But who owned and who directed such horses? Much later it was into the same regions, at various times and in various areas along the corridor, like beads on a necklace, that the Turkic-speaking peoples became established. Thus the nature of the country, and the way of life that it would sustain, notwithstanding modern and mostly misguided attempts to impose agriculture on the region, selected the peoples who could successfully occupy and move through such territory.

The migrations of language and plants are not directly connected. The ecology of the land has molded, at different times with different selective pressures, the distributions we now observe. Relating the patterns of language, ethnic groups, and seed dispersal becomes more complicated the more that is discovered (Shouse 2001).

Pack animals

There is a magnificent carved head of a horse about 15,000 years old from the French Pyrenees, but it is unlikely that horses were actually domesticated by then (Drower 1969). However, horses were certainly present in significant numbers in western Europe at that time and were probably exploited as an esteemed source of meat (Sherratt 2004). They were carefully observed, as is evident from the superb depictions in the paintings on the walls of the caves of Chauvet-Pont-d'Arc 30,000–33,000 years ago and at Lascaux about 15,000 years ago in southern France.

One of the earliest evidences of domestication comes from the remains of a horse, about 6,000 years old, from the archaeological site of Dereivka in the Ukraine. Tooth wear suggests that the horse had gnawed on its bit for much of its life; it was both domesticated and closely controlled. Contemporary experiments demonstrate that horses can be broken with bits made of wood or leather, but because of their perishability, neither kind is easy to find in the archaeological record. There is sounder archaeological evidence for the domestication of horses in and around the steppes to the north of the Black Sea (Drower 1969, Diamond 1991). Sophisticated horse-drawn chariots, generally associated with processional or military use (Sherratt 1984, 2004), were a feature of warfare in the Bronze Age. Therefore, although the development of technology about 5,000 years ago made bronze bits possible, and archaeological finds irrefutable, the earlier existence of domesticated horses is perfectly acceptable. The Ukrainian archaeological find is consistent with the analysis of mitochondrial DNA, which seems to imply that wild horses were tamed in many different places in Eurasia about 6,000 years ago. Of 191 domestic horses, 32 different maternal lines were revealed, suggesting that a large number of distinct female lineages must have been taken from the wild over a relatively short time (Vila et al. 2001). Horse use would soon have evolved into the separate disciplines of pulling, pack carrying, and riding.

Thus the first attempts at domestication may have begun, probably by capturing a mare with foal, as the first cities began to arise in the Middle East. The rapid and widespread domestication of the horse can be compared with the invention and spread of the motorcar. In any far-distant, future archaeological investigation, each would appear to have coincided with rapid acceptance, dissemination, and specialization.

By comparison with those of the camel, the teeth and guts of the horse do little or no damage to an apple pip. Indeed, they may enhance the potential for germination by cleaning and scarifying the seed. The horse, after domestication under human control, might therefore scatter both seed and a rich accompanying compost along the ancient east–west trackways. At oases, the horse might be tethered or hobbled. But unlike some other animals, a restrained horse is neither inhibited from defecating nor treading into its own droppings, much to the irritation of its handler. In the softer soil of the oases, its sharp hooves would have driven an apple seed into the turf. Iron horseshoes, which would have assisted such penetration, were not in common use until the Middle Ages, but the early horses of the steppe, the Tarpan and the Przewalski, had relatively narrow hooves. The wide "elephant's feet" of the heavy cavalry horse or shire horse, good for pulling the plow, are the results of much later selection.

In China, images of horses appear by the time of the Shang dynasty around 3,700 years ago. Chinese sources refer to the "heavenly horses" of the Fergana. The Fergana, or Ferghana, basin has for a very long time been a rich agricultural area, lying directly in the track of the westward migration of the sweet apple. The Chinese envy is understandable given that their lack of rich, nutritious, lowland grassland made it almost impossible for them to sustain populations of large grazing animals such as horses, cows, or sheep. It is worth noticing (Sherratt 2004) that the so-called Silk Road is called the Horse Road by the Chinese, reflecting their different priorities and aspirations. Similarly, the ass (*Equus asinus*), which is native to the savanna and scrub on the western desert margins, was also a later acquisition in the Orient, reinforcing the eastern accumulation of techniques from the West such as iconography, agricultural techniques, and domestic practices along the earliest routes of the Silk Roads.

Strangely, the horse appears to have been a relatively late arrival in ancient Greece and was certainly not there before 3,500 years ago. From the evidence of the first representative sculpture, these early Greek horses were small, scarcely larger than a modern polo pony, and were ridden without stirrups. Before the invention of mechanical transport, the horse and the ass were the chief beasts of burden, at least in Europe. The half-ass or onager (*Equus hemionus*) was used to draw chariots in early Mesopotamia but was not sustained as a beast of burden into later civilizations.

The camel is believed to have been domesticated about 5,000 years ago, possibly a little after the horse. The camel was soon found to have several advantages over the horse, ass, or onager: in good condition it could carry up to a 250-kg (¼-ton) load on its back and did not require iron shoes on hard ground or roads. The omnivorous camel eats carrion; indeed, its diet is legendarily untidy compared to the relatively fastidious horse's regime of grass, oats, and beans. As a beast of burden, the camel in historic time penetrated farther into Europe than is commonly supposed, for camel bones have been found in the Roman ruins at Vindonissa in present-day Switzerland (Bulliet 1975). It is also interesting to note that a late 18th century illustration of a caravan in the extreme west of what is now Turkey shows single-humped camels, horses, and donkeys, separately corralled (Plate 32).

It is commonly supposed that camels cannot be used to draw wheeled vehicles. Therefore, it is argued, the presence of cart or wheel fragments in grave goods presumes the use of horses. This is not altogether true, because camels, although intolerant of shafts, can be trained to pull carts on loose tethers (Bulliet 1975). Great care should thus be taken in the interpretation of the presence or absence of wheels or chariot parts in archaeological remains. Absence, too, need not suggest lack of knowledge of the technology. Because of the rocky and mountainous natures of their countries, both Greece and Japan have effectively disinvented the wheel for transport within the historic period. Moreover, on soft, sandy soils or snow and ice, sleds of various forms were used, and these are less likely to be preserved in archaeological remains.

Ancient apple names

Archaeological evidence for the collecting of apples from the wild in Europe can be found in Neolithic (about 11,200 years old) and Bronze Age (about 4,500 years old) sites throughout Europe (Hopf 1973, Schweingruber 1979, Jacomet 2005), and that for apple cultivation as early as 3,000 years ago in Israel (Zohary and Hopf 2000). Almost certainly from their size, however, these apples were the European *Malus sylvestris* (Plate 3; Villaret–von Rochow 1969), which includes *M. orientalis* of the Caucasus Mountains and Iran. In the era of recorded language that followed the often technologically advanced but illiterate peoples, the question of the name of the apple must be addressed. It also has to be asked when early peoples made the transition from collecting and using whatever small local crab was available and turned to imported, large, sweet apples from the Tian Shan. The position is very confused and often contradictory, and every tiny clue that can be scavenged from the very earliest written records must be evaluated.

Mesopotamia, literally the land between the rivers, the Tigris and Euphrates, is probably the oldest center of organized horticulture in the world (Map 8; Gopher et al. 2001). The first recorded parks were established in Assyria (roughly modern-day Syria and part of Iraq and Turkey) about 3,100 years ago, and the Hanging Gardens of Babylon, one of the Seven Wonders of the World, were created about 2,600 years ago. The Persian empire, which overran Mesopotamia, was at its height during the reign of Cyrus the Great (about 585 to about 529 B.C.) when his empire stretched from Greece to India. Under this imperial rule, agriculture and horticulture reached very high levels of sophistication.

The vernacular names given to the apple in the Middle East and Europe are many and generally confusing. But if the plethora of names there causes confusion, the problem pales into insignificance when compared to the early history of the fruit in China (Chapter 5).

The first use of words that may refer to the apple is a contentious subject but nothing to compare with the babel that was to follow. From the Fara period in early Mesopotamia (about 4,500 years ago), there are cuneiform texts with the word *hashur,* translated as apples in Akkadian dictionaries. There seems to be an etymological connection with the Syriac *hazura* (Postgate 1987). In these texts it is also recorded that *hashur* were available not only fresh but also dried and kept on strings. This written evidence is supported by the archaeological evidence that strings of dried apple rings were found in the Early Dynastic tomb (about 4,500 years ago) of Queen Pu-abi at Ur in Sumer (near present-day Baghdad in Iraq). But it is not yet possible to determine from DNA analysis whether these are referable to *Malus pumila* or another species. However, even allowing for contraction upon drying, their size, an average of 15 mm (less than ⅝ inch; J. M. Renfrew 1987) in diameter, suggests they were not *M. pumila*. It is worth noting here that the habit of drying apples on strings (Chapter 7), to preserve them from frost or insect or fungal damage and yet maintain their food value, was widespread in Britain and the United States until the turn of the 20th century. Drying apples is still common in the extreme climates of all the countries of eastern Europe and Inner and Central Asia (Plate 30).

The word for apple is well rooted in the ancient European language branches (Gamkrelidze and Ivanov 1984), including Celtic, Germanic, and Balto-Slavic. The suggested root is *ablu, ab(a)lo*, and for Germanic, *aplu* or *ap(a)*. In pre-Roman northwestern Europe, including Britain, employing that blanket but less than precise ethnic term Celtic (Cunliffe 2001), the apple is *abhall* or *abhal*. In Welsh it is *avall*. In Armoric (northwestern Gaul or Brittany), apple is *afall* or *avall;* in ancient Cornish, *aval* and *avel;* and in Gaelic, *ubhail*. In pre-Roman times in Britain, the mythical island center of civilization was called Ynys Avallach or Ynys Avallon.

Apfelbaum is the name for the apple tree in German; the fruit is *Opfel* or *Apfel*. In Old Russian apple is *jabl'ko*, in Russian *jabloko*. Apple is *tappuah* in modern Hebrew (which word is also used to denote the tree, but an important caveat is discussed later). The apple tree is called *melea* in Greek, but that is also the word for melon, resulting in some confusion. In Latin, again confusingly, apple is both *malus* (*malum*) and *pomum*. The Romans worshipped a goddess of fruit and called her Pomona, but their common word for the apple was *malus*. In addition, the word has been bedeviled by the close conjunction with many *mal* words with different meanings. The boy Miles, in Benjamin Britten's *The Turn of the Screw*, replies to his new governess with the hoary chant of generations of schoolboys, "Malo: I would rather be / Malo: in an apple-tree / Malo: than a naughty boy / Malo: in adversity." In setting to music three meanings of the *mal*—apple, evil or wickedness, and "I would rather be"—Britten might have included, to add to the confusion, Latin words with *mal* in the sense of mast and of jaw.

In the Bible, Genesis 3.6 records, "And when the woman saw that the tree was good for food, and that it was pleasant to the eyes, and a tree to be desired to make one wise, she took the fruit thereof, and did eat, and gave also unto her husband with her; and he did eat." Joshua 17.8 reads, "Now Manasseh had the land of Tappuah: but Tappuah on the border of Manasseh belonged to the children of Ephraim." Songs 2.3 and 2.5 has "As the apple tree among the trees of the wood, so is my beloved among the sons. I sat down under his shadow with great delight, and his fruit was sweet to my taste. Comfort me with apples." In Joel 1.12 there is "The vine is dried up, and the fig tree languisheth; the pomegranate tree, the palm tree also, and the apple tree, even all the other trees of the field, are withered."

It is universally agreed that the ancient Hebrew word *tappuah* should never have been translated into "apple." The general area of Palestine is much too dry and hot for general apple cultivation, and only comparatively recently have highly selected, low-chill varieties been introduced. The question of the identity of *tap-puah* has led to suggestions such as the orange, pomelo (another citrus fruit), quince, apricot, pomegranate, and even banana (de Witt 2000, Palter 2002), several of which are unacceptable on historical or botanical grounds. Nonetheless, the apple myth remains.

There is archaeological evidence for the consumption of apples of one species or another as far back as the transition from the Neolithic into the Bronze Age about 4,500 years ago. The evidence suggests that these were *Malus sylvestris,* more likely those once called *M. orientalis.* The fruits of this species are usually only 2–3 cm (¾–1¼ inches) across, rarely reaching 4 cm (1⅝ inches), and those identified as *M. orientalis* are always bitter and have to be dried or cooked to render their tannin-bitter flesh edible (Wiltshire 1995, Renard et al. 2001).

The evidence from the tomb of Queen Pu-abi remains equivocal, but it suggests that apples of some sort were being shipped down the great rivers of Mesopotamia, probably from the Elburz Mountains south of the Caspian Sea, the Zagros Mountains along the Iran-Iraq border, and Anti-Taurus Mountains of eastern Turkey, all to the north. Like Palestine, Mesopotamia on the whole would have been far too hot for apple growing. Cuneiform writing about 3,800 years old confirms the bringing of apples down the great Mesopotamian waterways. Although highly suggestive, this evidence is not absolutely definitive in favor of *Malus pumila.*

But from the records of Alexander the Great, just before his death in 323 B.C., can be gleaned a small and perhaps significant clue. Training his battle fleet on the Euphrates, just outside Babylon, near Baghdad in present-day Iraq, Alexander put the crews to the test in simulated sea battles. Crew pelted crew with apples from the decks of the royal fleet (Lane-Fox 1973). It would have served little purpose for the warriors to have used any form of crab apple barely 2.5 cm (1 inch) across, even propelled by a muscular Macedonian arm. But to receive the full force of a large Tian Shan apple, perhaps one of the hard-skinned, phase 3 type, would have been a salutary educational experience!

If Sumerian cuneiform texts and archaeological remains suggest sophisticated cultivation of the apple more than 4,000 years ago, are there similar pieces of evidence that suggest the apple's early use nearer to its place of origin? The evidence from Chinese sources is reviewed in Chapters 4 and 5.

CHAPTER 4

Apples and grafting

EARLY DOMESTIC APPLES were grown from seeds and produced a range of types as they rarely "come true." As the apple was brought west, it came into contact with peoples practiced in grafting, so grafted clones would ensure a uniform, reliable crop (cuttings are generally difficult to strike). Grafting of grapes was perhaps understood about 3,800 years ago, and there is good evidence of grafting 2,500 years ago in Persia. By the Roman period, grafting was almost as sophisticated as it is today. Grafting as a technique may have arisen independently in China, for mulberries used in silk production. Dwarfed apple trees were known 2,300 years ago, and Paradise rootstocks were probably brought to the West, ultimately from China, via Armenia.

THERE IS EVIDENCE for the regular human occupation along the east–west trade routes possibly as early as 10,000 years ago (Map 4) and certainly 7,000 years ago. But what was happening to the apples of the Tian Shan?

It is now becoming clear that the first signs of developing agriculture, as opposed to mobile hunting-fishing-gathering, can be detected somewhere in the region of what are now southeastern Turkey, northern Syria, and Iraq (Map 8). This region includes areas where that essential nutritional combination of the grain and the legume occur naturally, namely, the wild ancestors of wheat (*Triticum aestivum*), barley (*Hordeum vulgare*), chickpea (*Cicer arietinum*), and lentil (*Lens culinaris*). And the dates of this transition to the Neolithic are being steadily pushed back, approaching 13,000 years ago (Lev-Yadun et al. 2000, Gopher et al. 2001). Thus, as the apple (*Malus pumila*) began to be moved west, this new crop would have been brought into areas where peoples had been familiar with agricultural techniques for several thousand years, including the cultivation of tree crops such as figs. Therefore, they would be expected to be receptive to a new source of fruits. The apple had the enormous advantage that it could be harvested and consumed easily. Hulling, peeling, grinding, washing, cooking, infusing, fermenting, and the washing out of toxic substances were unnecessary. Most apples, when kept under the best conditions, improve in quality. As John Worlidge wrote in 1669, "not a day in the year but they may be had, and not of the worst." Few food sources are as convenient. This welcome economic migrant was on its way to conquering the West.

The apple must have advanced west for several thousand years exclusively in the form of seedlings from randomly open-pollinated flowers. Despite the fact that the transported apples would have been selections from the larger, sweeter, juicier representatives of the fruit forest, modern experience suggests that the progeny would have varied, from the edible but also to the impossibly astringent (Brown and Maloney 2003: pl. 3.2).

Since apples are self-incompatible, it would soon have become apparent that they do not usually come true from seed ('Blenheim Orange' appears to be a partial exception, setting some viable seed—as a triploid it would not be expected to form seed in the normal way through fusion of viable gametes). Given the high

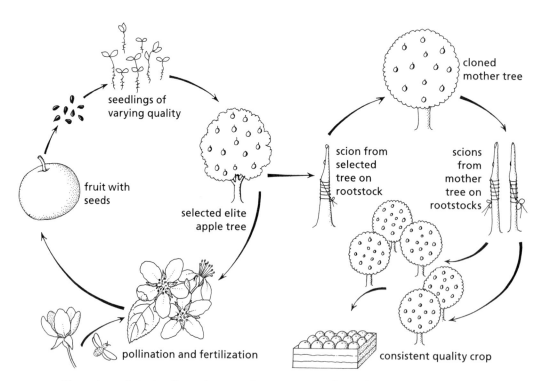

The nature of cloning through vegetative propagation by grafting versus the variability resulting from sexual reproduction. Drawing by Rosemary Wise.

degree of polymorphism in *Malus pumila*, offspring frequently do not resemble their parents.

The high rate of failure to get the tastiest apples by planting seeds would have encouraged the early farmers to propagate plants through cuttings or, possibly, layerings, but the success rate would have been negligible for such material. The relationship between random, open pollination, seedling diversity, and the development of grafting (a form of cloning through vegetative propagation) is illustrated in the accompanying diagram, but for many hundreds if not thousands of years the true nature of what was happening was not understood. And the mechanism of the success of the graft union (Plate 36), given the complexity of the meristematic fusion, is still not wholly understood.

It will probably never be known who invented grafting, but two nearly simultaneous developments are possible to imagine, neither directly connected with the apple. Having in their immensely rich flora far more fruits than were available in the West, the Chinese laid no great store by the apple. Many fruits, such as the much venerated peach (*Prunus persica*), would have self-seeded and more or less regularly bred true. Moreover, the apple comes from a region in the far northwest of China, Xinjiang Uygur, over which successive Chinese dynasties had, at best, only vestigial control over much of history.

Conventional grafting, what has been called "instant domestication" (Zohary and Hopf 2000), is possible in skilled hands between species of practically any of the genera of Pyreae. The Chinese regularly use species of *Cotoneaster* as rootstocks for apples, while the quince (*Cydonia oblonga*) is routinely used as a commercial rootstock for pears (*Pyrus*). To achieve a degree of dwarfing, pears were grafted onto hawthorns (*Crataegus*), certainly until medieval times (Amherst 1895). All these cross-compatibilities suggest a fairly recent evolutionary origin for the group as a whole. A comprehensive review of the earlier literature can be found in Roberts (1949), and the contemporary scene is fully covered by Wertheim and Webster (2003).

Apples, as well as most other Pyreae, as explained before, cannot be easily propagated by cuttings or layerings. There are bitter, semi-soluble phenolics in the young bark that bind readily to any protein, hence the use of such bark to make leather from animal skins. The phenolics, called tannins, link the proteins together. In plants, these tannins appear to help protect first-year shoots against browsing. As the bark is broken in taking a cutting, these tannins also seem to inhibit the development of adventitious roots that would grow from the base of such a cutting. Successful cuttings can be struck from a very few exceptional apples such as 'Burr Knot' or 'Burr Knot Lascelles', 'Chiloe', 'Sheep's Snout', 'Irish Pitcher', and 'Oslin' (or 'Orgeline' of Bunyard and Thomas 1906; Bunyard 1920) but these are all of limited commercial value. 'Tom Putt', dating from the late 1700s, will usually root, along with the ancient cider cultivar 'Genet Moyle' (*moyle*, mule, or hybrid).

Apple trees that develop burr knots—small, corky, sometimes pointed protuberances usually near the base of a stem—will usually root from cuttings. No less an authority than Charles Darwin (1845), in his *Journal of Researches* from the voyage of the *Beagle*, recorded on 4 February 1835 that apples apparently grown from cuttings flourished over large areas of the island of Chiloé off the coast of Chile: "the inhabitants possess a marvellously short method of making an orchard. At the lower part of almost every branch, small, conical, brown, wrinkled points project: these are always ready to change into roots, as may sometimes be seen, where any mud has been accidentally splashed against the tree." He also noted extensive tracts of self-sown seedlings there. Fruit trees growing on their own roots have some advantages and their adherents. The latest techniques such as growth promoters and bottom heat (Webster and Wertheim 2003) can yield some success, but these techniques were not available to the ancients.

Grafting and the apple's move westward

The large, sweet apple, as it spread from the Tian Shan westward along the great trade routes, would have scattered its diverse, outcrossed progeny, the products of very many pollination events. Seed material might also have moved back into the fruit forest as caravanners foraged. From time to time, the wild animals of the forest, too, moved to neighboring pockets of forest. Thousands of miles of desert trackway, particularly around the oases, would have served as a crude selecting ground with a constant stream of travelers and their animals unconsciously noting, evaluating, and dispersing the seeds of certain fruits.

Human beings mostly chew apples down to but not including the fibrous placental tissue, or core, and the bitter pips, or seeds (see page 27). Likewise, birds will eviscerate an apple (see page 31), but because of the cyanide content of the apple pips, they tend to leave the seeds undamaged and also still within the placental tissue. Leaving the seeds within the core would have been of limited value in disseminating potential seedlings (Chapter 1). Horses, and bears and some of the minor inhabitants of the forest such as wild boar, deer, and badgers, are better at separating, partly scarifying, and distributing the seeds. The winter climate of Inner and Central Asia, and indeed the more continental areas of Turkey and Spain, would have ensured adequate cold-chill and subsequent germination.

There is little stimulus for the development of grafting as long as a there is a rich diversity of fecund seedling growth. Every roadside fruit stall in Uzbekistan, Kyrgyzstan, and Kazakhstan (Plate 27) displays a cornucopia of apples in season. But enquiries as to the name of the cultivar will be met with confusion and the explanation that any particularly choice apple comes from somewhere up an adjacent valley. Why be concerned with time-wasting propagation methods when diversity prevails, and when a favored apple tree dies there are so many to take its place? The losers in this race for genetic superiority are converted to firewood (Plate 26).

But as the apple spread into the drier lowlands of what is present-day Turkmenistan, Iran, and Iraq, germination would have become rarer. Probably only around the oases and other fertile areas where seedlings might flourish, and horses might graze and browse, would small populations of apples and other seedlings have emerged. Horses would have driven the seeds into the soil with their hooves; dung beetles may have assisted in the burial.

The technique of grafting might have been invented as a result of the casual observation by field workers of the natural grafts that can occasionally be found between adjacent plants of the same species (Plate 36). Natural grafts between root systems and, more rarely, aboveground parts have been seen in *Malus pumila*

Natural pressure grafting, which might have taken place when a temporary bender shelter was made by field workers. Drawing by Rosemary Wise.

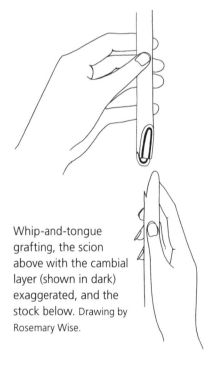

Whip-and-tongue grafting, the scion above with the cambial layer (shown in dark) exaggerated, and the stock below. Drawing by Rosemary Wise.

Cleft grafting from Leonard Mascall's *A Booke of the Arte and Maner, Howe to Plante and Graffe All Sortes of Trees . . .* (1572). The curiously bandaged appearance of the cleft-grafted stock, with adjustments for the perspective and scale, is explained: "The usual cover for protecting the [s]cions, is clay well tempered, and mixed with horse-dung; an excellent substitute, which may be kept ready for use when a little softened by heat, is a mixture of equal parts of tallow, bees-wax and rosin, spread on strips of linen or paper six inches long and about two inches wide [15 by 5 cm]; one of these strips wrapped around each stock, so as to completely cover the fissure at the sides and in the end" (Coxe 1817). The addition of horse dung, though not widely practiced now, would probably have enhanced the quality of the graft by adding plant growth-promoting auxin analogs from the urine.

in the Tian Shan (Dzhangaliev 2003). It is just possible that this information was remembered and transferred by migrant peoples as the apple moved out of the mountains westward, when conventional cuttings and seed propagation were found unsatisfactory. It will probably never be known which method of grafting was first developed: cleft grafting (Plate 33) or whip-and-tongue grafting.

Another possibility is that where agricultural workers were engaged in long-term field clearance or tillage, they might have made overnight shelters, using benders, basic frames of flexible branches used as hoops and covered with waterproof material. Temporary shelters of all types are still a feature of agricultural areas throughout Inner and Central Asia. If these benders were made of cut stems and already rooted material flexible enough to be bent into a curve, as in species of ash (*Fraxinus*) or osier (*Salix*), and strapped tightly together with withy stems, the compatible species might have both rooted and formed pressure grafts.

Archaeological evidence for grafting

In 1934, French archaeologists uncovered the great palace at Mari on the middle Euphrates some distance to the north of Ur in what is now Syria. The whole area had been highly civilized beginning at least 4,300 years ago. Much of the evidence for this is inscribed on palm-sized tablets of unfired clay on which information on domestic, state, and military matters is recorded in cuneiform. On one occasion, kitchen staff received a shipment of 100 liters or quarts of apples and 10 of medlars from upriver. Presumably, these had been shipped down the Euphrates from the mountain slopes of the cooler Elburz, Zagros, and Anti-Taurus Mountains to the north. But crucially, there is also tantalizing evidence that grafting of grape vines (*Vitis vinifera*) might have been understood (Lion 1992). Cuneiform writing is not easily translated, and the interpretations are extremely tentative. Nevertheless, the Babylonians were sophisticated and observant farmers. The date, about 3,800 years ago (according to palace dating, which adds together and integrates all the known dates of kings, emperors, civil servants, and historical events in an attempt at a continuous chronology and is now thought now to be accurate to within 60 years, plus or minus), would be consistent with movements along the great trade routes to the north.

Knowledge of and samples of other fruits to which such techniques could be transferred must also have filtered north–south. Possibly, the farmers of Mesopotamia were already troubled by salt accumulation as a result of irrigation, in dry heat, over several thousand years. They may have realized that some vines of indifferent fruit quality were nevertheless tolerant of salt contamination in the

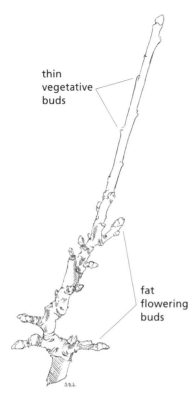

thin
vegetative
buds

fat
flowering
buds

A "lance" above, the first-year wood growth with thin vegetative buds, used as a scion, and older wood below with fat flower buds. Drawing by Sarah Juniper.

soil. These became the rootstock plants. Other vines, as it were the 'Cabernet Sauvignon' of the period and region, were salt sensitive but as scions (bud-bearing shoots) could be grafted onto the salt-tolerant rootstock.

By the reign of Cyrus the Great in 558–529 B.C., the sophisticated and horticulturally skilled Persian orchardists would certainly have noted the natural grafting of adjacent trees and shrubs. The building of formal gardens, by then widespread, with their trees and shrubs trained through pleaching, would have encouraged natural and visible pressure grafts. In the ebb and flow of armies, particularly those of Alexander the Great two centuries later, such lore would have passed to Macedonia and Greece by the forced relocation of skilled slaves, a feature of Alexander's conquests, and the movement of travelers. Alexander was a noted cavalry commander (Lane-Fox 1973), and the movement of men and horses could not have failed to move fruit seeds of all kinds. Is it possible, too, that some of this knowledge, extracted from Mesopotamia, was passed to the Celtic peoples, whose influence 2,500–2,100 years ago extended from what is now southern and central Turkey through Greece way beyond to Ireland and to the Shetland Islands off northern Britain (Map 7)?

Darius I, king of Persia 522–486 B.C., wrote a letter to one of his satraps, Gadatas, in Magnesia, an ancient historical region now part of western Turkey (Map 8; Meiggs and Lewis 1989). The letter is a stone carving that was part of a building, possibly a temple, and is now in the Louvre in Paris. In it, Darius congratulates Gadatas for cultivating crops and fruit trees from Syria in western Asia Minor; Darius himself may have been in western Asia at the time. Such a text hints strongly at grafting, and the selection of quality fruit, but again the evidence is not quite definitive. It is interesting to note also from that at this time Magnesia must have been well within the eastern sphere of influence of the Celtic peoples.

The ancient Greeks built extensive orchards, which can be interpreted, with some reservations, as a product of grafting techniques. The Romans learned the techniques partly from the Greeks and partly from the Syrians. It is recorded in Roman writings, "The Syrian slaves brought with them, along with other oriental corruptions of the senses, also the oriental refined ability to care for animals and plants. Like the castration, circumcision and the production of bastards, it was common in the early days to prune the trees and mix the fruit varieties by grafting" (Hehn 1902). This slight sense of revulsion, that there is something wrong and unwholesome in the artificial mixing of two or more plants, is widespread. The practice of joining together two plants of different origins is forbidden under Hebraic law (Leviticus 19.19), it was condemned in the 16th century by the botanist Jean Ruel and reviled by "Johnny Appleseed" Chapman, but none of this stopped anybody using the technique when and where it suited them.

By the time of Cato (234–149 B.C.) there were instructions on how to generate new trees from living trees using bored pots or baskets filled with earth (Hehn 1902), in other words, a form of air layering, which will work for a small number of apple cultivars. Moving forward to the time of Pliny the Elder (A.D. 23–79), there is evidence of a substantial number of distinctive cultivars of apples, perhaps as many as 100. There is even evidence that Pliny experimented with what we would now call family trees, the usually unsatisfactory practice of grafting several cultivars onto one rootstock. Pliny writes of, among others, the 'Scantian Apple' (presumably a phase 3 type since it was stored in casks), the 'Little Greek', the 'Syrian Red', the 'Must Apple' or 'Honey Apple', for its rapid ripening and its flavor (obviously a phase 1), the 'Blood Apple', for its color, and the 'Sceptian', named after its discoverer, a freed slave. 'Annurco' was also cited under the original name of 'Orbiculata', locally called 'Orcola', which is still widely grown in the Campania of southern Italy (Pliny 1967). It is possible, though contentious, that the Roman cultivars 'Decio' (thought to date from the time of Attila, about A.D. 450, and perhaps brought west by the Roman general Ezio; Grassi et al. 1998) and 'Court Pendu Plat' still survive and are grown in Britain and elsewhere today. 'Decio' has certainly been grown for at least 500 years (Morgan 1993), and there is no fundamental genetic reason why these cultivars should not have survived since Roman times, although many well-known cultivars of the past have been overwhelmed by accumulating viruses and the like.

A Roman supply ship went down off Palma during the invasion of Majorca in the Balearic Islands of the western Mediterranean in 125 B.C. (Map 7). In the hold were amphorae of olive oil, whole olives, almonds, wine, and in addition, rammed into a lump of clay and wrapped with a rag, a bundle of vine cuttings (Damien Cerda Juan pers. comm. 1995). Through his intelligence services, the invading Roman commander was obviously aware of the inadequate horticulture of the island and the absence of any sophisticated provisions. It is beyond doubt

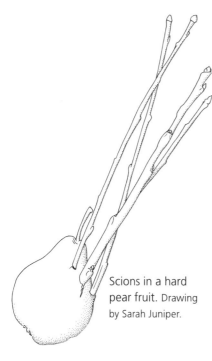

Scions in a hard pear fruit. Drawing by Sarah Juniper.

that the Romans and earlier civilizations were aware of the other effective method of transferring scion wood safely over vast distances, namely, thrusting lengths of scion wood deep into a hard fruit such as quince or pear, or into a root vegetable such as beet. Probably in the rough movement of a ship, the clay lump method was preferable.

A few years later, through the Roman poet and philosopher Lucretius (about 90 to about 50 B.C.), there is written evidence of grafting (Lucretius 1907). The techniques were developed to such a level of sophistication by the Romans that there is little that a modern-day horticulturist could have taught the 1st century A.D. writers Columella (Forster and Heffner 1979), author of *De Re Rustica,* or Pliny the Elder on even the finer points of grafting, budding, dwarfing rootstocks, and other techniques. Indeed, the Romans laid a mosaic at Saint-Romain-en-Gal in southern France, dating from about the first half of the 3rd century (Pelletier 1975), in which the whole of the orchard year is depicted, including planting, cleft grafting (Plate 33), and pruning through to olive, grape, and apple harvesting, and the preparation of olive oil, wine, and cider (Plate 34).

Grafting in the Middle Ages and after

Following the withdrawal of the Roman legions from Britain about A.D. 450, there were many hundreds of largely unrecorded years, the so-called Dark Ages. About 1,000 years passed from the end of Roman rule till the first irrefutable written evidence of grafting in medieval England. But on vigorous rooting stocks, such a period of time need only represent five successive events and five learning occasions. Incontrovertible archaeological supporting evidence will almost certainly never be uncovered, but it seems reasonable that such an important horticultural technique as grafting, applicable to so many crops for so many purposes, survived in rural practice if not in texts through the Dark Ages. Indeed, even in the literary darkness there are occasional tiny glimpses of orchard activity. Janson (1996) lists all the scraps of evidence from A.D. 360 to 1340, leaving no doubt that at least orchards and grafting survived the breakdown of central authority.

In his translation of Pope Gregory the Great's *Regula Pastoralis,* King Alfred wrote about the year 897, "Sio halige gesomnung Godes folces, thaet eardath on aeppeltunum, thonne hie wel begath hira plantan & hiera impan, oth hie fol-weaxne beoth," to the holy congregation of God's people which live in apple orchards, where they well tend their plants and their grafts until they are fully grown. The cluster of British place-names with *apple-* prefix and strong Celtic or Saxon associations (Chapter 5), suggests that groves of sweet apple trees were revered. Even in the Dark Ages, it seems unlikely that the small, straggly, native crab with its virtually inedible, astringent fruits would have been granted so striking a memorial.

The fact that "graffing" is frequently commented on in texts in various languages in northwestern Europe from the late medieval period onward (Janson 1996) suggests that the technique was widespread and sustained. ("Graffing," a rendering soon widely superseded by "grafting," persisted in the written form at least until Nehemiah Grew's book of 1675.) Walking through the local countryside about the year 1125, William of Malmesbury, librarian of the Abbey of Malmesbury in Wiltshire, England, a brilliant son of a Norman father and a Saxon mother, commented in his *Gesta Pontificum Anglorum,* the achievements of the bishops of England, "The land everywhere bears crops in abundance, and is fertile in fruit, whether naturally or thanks to the skill of the farmer; the conditions are such as to encourage even the most bored slacker to take kindly to labor, seeing that the response will be a hundredfold. You may see the public highways decked with fruit trees [*arboribus pomiferis*] thanks not to art and industry but to the very nature of the soil. The land produces fruit spontaneously, fruit that is far superior to all other in flavour and beauty. Many do not wither away by the end of the year, but do their duty by their masters until replacements come along" (Winterbottom and Thomson 2004).

With modern knowledge and hindsight, it can be concluded that not only had the people of the Gloucestershire-Wiltshire border country retained the skills of grafting ("thanks to the skill of the farmer") and that some cold winters in the immediate past had led to the spontaneous germination of seedlings, but also that some trees held their fruit for a long time (Plate 35) or in cool storage, or both, surviving as good fruit until apples came again.

Michael Hennerty (in Kennedy 1997) has unearthed evidence of apple cultivation in Ireland, wherein there is a glowing reference to the MacCann clan leader who died in Armagh in 1155 and was remembered "for the strong drink made for his clan from apples grown in his own orchards." Given the long life of orchards and the persistence of rural techniques, as recorded by William of Malmesbury, it seems possible that this brief sketch records the existence of grafting and the assembly of very specific cultivars perhaps before and certainly only a few years after the Norman Conquest. Could it be that since neither the Romans, nor to any great extent the post-Roman invaders of northwestern Europe, penetrated and attempted to subdue Ireland, an early Celtic tradition of apple growing had persisted?

King Alfred's translation from the Latin into Old English, the first written mention of the technique of grafting, would scarcely have been readable by Chaucer. The first real vernacular horticultural mention of grafting in the evolving modern English language is probably to be found in that strange 15th century document *The Feate of Gardening* by Maister John Gardener, or Gardyner (manuscript in Trinity College, Cambridge; see Amherst 1894, 1895, Harvey 1985), whose real name may have been John de Wyndesores or John de Standerwyck. All the possible names have their advocates, but whatever the author's true identity, it seems likely that it was the work of a leading master gardener, possibly at the royal palace of Westminster or at Windsor Castle.

The exact date when *The Feate of Gardening* was written is also debated, most authorities suggesting a date around 1440 but some putting it as early as the mid-1300s. The two original versions, a complete one in Trinity College, Cambridge, and an incomplete version, the so-called Loscombe (Wellcome Historical Medical Library manuscript 406), are now only a handful of manuscript pages in doggerel verse. They do not seem ever to have been published, so they cannot have had much influence on the horticulture of the time. However, the work is entirely practical and shows no sign of borrowing from other works (Harvey 1985). The text has a short introduction, leading to sections on trees, grafting (using the rendering "graff"), viticulture, the onion family, coleworts (brassicas), parsley, herbs, and concluding with saffron.

The first published and widely distributed text with grafting accurately described in English was John Fitzherbert's *The Boke of Husbandry* of 1523. This text is so accurate in every detail

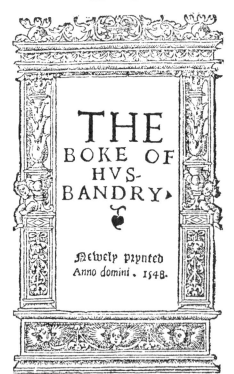

THE
BOKE OF
HVS-
BANDRY.

Newely prynted
Anno domini . 1548.

Title page of a later edition of John Fitzherbert's *The Boke of Husbandry.*

and so thoroughly instructive as to suggest that grafting techniques were both widespread and well understood:

❡ Howe to graffe.

Thou muste gette thy graffes of the faireſt lanſes, that thou canſte fynde on the tree, and ſee it haue a good knotte or iopncte, and an euen. Than take thy ſawe, and ſawe into thy crabbetree, in a faire plapne place, pare it euen with thy knyfe, and than cleane the ſtocke with thy greate knyfe and thy maſſet, and ſette in a wedge, and open the ſtocke, accordynge to the thyckeneſſe of thy graffe, than take thy ſmalle ſharpe knyfe, and cutte the graffe on bothe ſydes in the iopncte, But paſſe not the myddes therof for nothyng, and ſet the inner ſyde, that ſhall be ſet into the ſtocke, be a lpttell thynner than the vtter ſyde, and the neither popncte of the graffe the thynner: than proferre thy graffe into the ſtocke, and if it go not cloſe, than cut the graffe or the ſtocke, tyll they cloſe cleane, that thou canſte not put the edge of thy knyfe on neyther ſyde betwene the ſtocke and the graffe, and ſette theim ſo, that the toppes of the graffe bende a lpttell outewarde, and ſee that the woodde of the graffe be ſet mete with the woodde of the ſtocke, and the ſappe of the ſtocke maie renne ſtreyght and euen with the ſappe of the graffe, for the barke of the graffe is neuer ſo thicke as the barke of the ſtockes And therfore thou mayſte not ſette the barke mete on the vtter ſyde, but on the inner ſyde: than pull awaye thy wedge: and it wyll ſtand muche faſter. Than take toughe clepe, lyke marlep, and lep it vppon the ſtocke head, and with thy fynger lape it cloſe vnto the graffe, and a lpttell vnder the head, to kepe it moyſte, and that no wynde come into the ſtocke at the cleauynge. Than take moſſe, and lape thera vpon, for chpnynge of the clape : than take a baſte of whpte wethy or elme, or halfe a bypcer, and bynd the moſſe, the clap, and the graffe togethcr, but be well ware, that thou breake not thy graffe, neptber in the claipng, nor in the bindpnge, and thou muſte ſet ſome thpnge by the graffe, that crowes, nor bprdes lpghte not vpon the graffe, and breake hpm, &c.

How to graft. Thou must get thy grafts of the fairest lances [the slim, upward-pointing, flat-budded, first-year wood suitable for scions] that thou canst find on the tree, and see it have a good knot or joint and an even [one]. Then take thy saw, and saw into thy crabtree in a fair plain place, pare it even with thy knife, and then clean the stock with thy great knife and thy mallet, and set in a wedge, and open the stock, according to the thickness of thy graft, then take thy small sharp knife, and cut the graft on both sides in the joint, but pass not the middle thereof for nothing, and set the inner side, that shall be set in to the stock, be a little thinner than the other side, and the neither [nether] point of the graft the thinner: then proffer thy graft into the stock, and if it go not close, then cut the graft or the stock, till they close clean[ly], that thou canst not put the edge of thy knife on neither side between the stock and the graft, and set them so, that the tops of the graft bend a little outward, and see that the wood of the graft be set meet with the wood of the stock, and the sap of the stock may run straight and even with the sap of the graft, for the bark of the graft is never so thick as the bark of the stocks. And therefore thou must not set the bark meet on the outer side, but on the inner side: then pull away thy wedge: and it will stand much faster. Then take tough clay, like marl, and lay it upon the stock head, and with thy finger lay it close into the graft, and a little under the head, to keep it moist, and that no wind come into the stock at the cleavage. Then take moss, and lay thereupon, for chining [to prevent the cracking or fissuring] of the clay: then take a paste of white withy of elm, or half a briar, and bind the moss, the clay, and the graft together, but be well aware, that thou break not thy graft, neither in the claying, nor in the binding, and thou must set some thing by the graft, that crows, nor birds light not upon the graft, and break him, etc.

On the continent of Europe, Johann Domitzer's influential *Ein Neues Pflantz-büchlin*, a new little plant book, published in Augsburg in Bavaria in 1531, gave a comprehensive account of grafting and was widely consulted well into the 17th century. In 1536, Jean Ruel, writing in Latin from Paris, frequently refers to the practice of grafting though he gives no description of grafting as such, discussing *miscella insitione . . . insitionis adulteries,* mixed insertions . . . adulterous insertions (note the prevalent discomfort with the practice of grafting, the mixing of plants by human intervention, a sentiment echoed today in the debate about genetic modification of crop plants). Ruel also noted (translation by Brendan McLaughlin), "whereas the earlier Roman writers such as Cloatius listed a fair number of apple varieties, later horticulturists far surpassed their predecessors through their skills in grafting." Such comments suggest that the horticultural technology from Persia via the Romans was known in medieval Europe, not only by word of mouth and by example but also from well-distributed literary sources, many of which are now lost.

Thomas Tusser (1557, 1573) had quite a lot to say in his strange, versified literary form on fruits. And by the time of Thomas Hill (1563) and Leonard Mascall (1572) it was possible in Britain to buy illustrated textbooks of quality in a recognizable, modern, instructive style. They were expensive, but like Domitzer and unlike Ruel, they were in the vernacular. They not only described but also depicted the techniques of grafting as well as the tools of the trade.

Shakespeare knew about grafting, as about almost everything else in his time. In *Coriolanus* (written between 1599 and 1608), he wrote in Act 2, Scene 1, "We have some old crab-trees here at home that will not be grafted to your relish." John Parkinson (1629) not only described more than 200 fruit cultivars, of which 75 were apples, but also grafting techniques and equipment in great detail. Entries in the account books of Sir Richard and Lady Lucy Reynelss of Forde for March 1645 (Gray 1995), "Robert Salter and William Salter for grafting 14 [pence]" and "To the boys for claying the grafts 06 [pence]," presumably indicate cleft grafting in which the exposed tissue, as explained before, was protected with clay. The day rates for labor are not given, but it may be assumed that the Salters received 7 pence per day for their work, and the boys (two of them?), 3 pence each.

The first written evidence of double working, in which a scion is grafted onto a stem that is itself grafted on a different rootstock, a technique apparently known to the Romans, would appear to be in Thomas Langford's (1681a) *Plain and Full Instructions to Raise All Sorts of Fruit-Trees:* "Mr. *Rea* [1665 and many later editions] judging the *Paradise-Apple* of somewhat slow growth in bringing forth a *Cyen* [scion], advises to *Graff* a *Paradise* on a *Crab-stock,* and the *Fruit* you would have on that *Paradise,* that the *Crab* might yield plenty of *juice* or *sap* to the *Paradise,* and the *Paradise* retard the growth of the *Apple planted* on it, so as to keep it *Dwarf.*" Jean de la Quintinye (1626–1688; Evelyn 1664, London and Wise 1699a, b), gardener to Louis XIV of France and certainly the most outstanding pomologist of the 17th century (called "father of the pruning art"), is recorded as using dwarfing (Paradise) rootstocks for both apple and pear in the famous gardens he developed at Versailles. Among other things, de La Quintinye gave advice on fruit storage in stores with walls 2 feet (0.6 m) thick, double-glazed windows, an entrance air lock, and even a cat flap for rat- and mouse-hunting felines.

Following on from, and plagiarizing to a considerable extent, both Langford and de La Quintinye was that colorful character Richard Bradley (1688–1732). Bradley succeeded in getting himself expelled both from the Royal Society of London and the chair of botany in Cambridge for questionable behavior. He seems never to have delivered any lectures in the latter post but had them printed as books, no doubt a much more profitable exercise. He published about 20 books from about 1717 onward, was no mean botanical artist, and did much to advance the "philosophical," technological, aspects of orchard growing.

Grafting in China

The Chinese, with their immense richness of fruit species, expressed no particular interest in the apple even though a portion of the Tian Shan lay within their present political borders (Simoons 1991). Xinjiang Uygur was to them a faraway region, not infrequently passing out of their control. What is more, it lay beyond a peculiarly hostile region, namely, the Gobi. Today, many Chinese are of the opinion that the apple, as typified by the Western cultivars that began to appear in their markets about 1860, is a foreign fruit. In any case, many of the very early Chinese cultivars were of a mealy texture, were accorded little respect, and did not keep long.

Nevertheless, there was another crop of immense value to the Chinese dynasties that lent itself to the development of special propagation techniques, including grafting. The early Chinese did not possess much in the way of precious goods—little gold and neither silver nor precious stones—with which to trade. Yet they required certain items from the West, such as cobalt from Persia (Carswell 2000a, b), lapis lazuli and emeralds from Afghanistan, and high-quality, large horses from the Fergana Valley in what is now mostly Uzbekistan (Map 5). For these they needed a currency with which to reward vassal or would-be friendly states on their long, open, embattled borders. To this end, from about 2,300 years ago, they developed a sophisticated silk industry (Fortune 1847). The silkworm (*Bombax mori*) feeds almost exclusively on the leaves of the indigenous white mulberry (*Morus alba*). The edible and much-prized dessert fruit, the black mulberry (*M. nigra*), is a barely adequate substitute.

Gao Jien (pers. comm. 1999) believes that the practice of grafting may have originated independently in China, not for fruit but for silk production. Direct written evidence prior to the Christian era does not now seem to exist. Like the unrelated Pyreae, neither species of mulberry roots readily from conventional cuttings. From a relatively early age, a mulberry tree develops long, pendulous branches that droop to the soil surface and, by self-layering, slowly extend the original tree. Tidy-minded gardeners have, for generations, either pruned off these inconvenient drooping limbs or scrupulously mown and cleaned underneath the lowering branches, thus frustrating the proliferation.

Among the various species of *Morus* there is a diversity of vegetative forms that might have stimulated the development of grafting techniques. Orchard trees most suitable for feeding the armies of silkworms would have been developed through this layering technique by cloning trees bearing leaves in high yield. On the other hand, the tiny heads of white fruitlets of *M. alba* were of no particular value to the Chinese and could have been ignored except in a very few localities (for example, Tajikistan) where they are still dried for winter consumption.

Qi Min Yao Shu, necessary skills for the masses, a book in 10 juan (a word originally meaning a roll on silk cloth, or strips of wood joined together and rolled into a drum, later coming to mean a chapter), was written by Jia Sixie, the prefect of Gaoyang, Hebei province, not far south of Beijing (Map 2), and completed sometime between A.D. 534 and 550. It deals with general agricultural techniques (plowing, harvesting, sowing), the growing grain crops, vegetables, fruit trees, other useful trees, animal husbandry, the making of alcoholic drinks, processing of foodstuffs into sauces and the like, and other basic skills useful to peasants. Its last juan deals with plants not occurring in China. In juan 5, section 45, concerning a range of vegetative methods of propagation, is recorded the following (translation by Stephen Haw): "To layer mulberries: during the first and second months [roughly equivalent to mid-February to mid-April], use a hooked stick to peg down lower branches into the earth. When leafy shoots have grown up to a height of several inches, earth them up with dry soil (if the soil is wet then they may rot). In the first month of the following year, cut them off, dig them up and transplant them." And in juan 4, section 37, on grafting pears:

Grafting is much quicker [than growing from seed]. The method of grafting is as follows: use *tang* or *du* [wild pears]. (If *tang* are used, the pears will be big with fine flesh [possibly *Pyrus phaeocarpa* Rehder or *P. betulifolia* Bunge]; *du* are not quite as good [possibly *P. calleryana* Decaisne or *P. ussuriensis* Maximowicz; there are not many other possibilities—fewer *Pyrus* than *Malus* species are native to China]. If mulberry is used, the pears will be very bad. If jujubes or pomegranates are used to graft pears on to, out of every ten grafts only one or two will take.) *Du* as thick as an arm or thicker are suitable for use as stocks. (Plant *du* in advance, grafting a year later. Stocks can also be planted when needed for grafting, but then if the *du* stock dies the scion cannot live.) Large *du* trees can have five branches grafted onto them, small ones three or two.

The best time to graft is when the leaves of the pear are just beginning to emerge from the buds. The latest time is when the flower buds are just about to open. First make hemp thread and bind about ten turns around the *du* stock, then saw it off 5 or 6 inches (13–15 cm) above the ground. (If the stock is not bound, then it is to be feared that the bark will split during grafting. If the *du* stock is left taller, the pear branches will flourish, but may break in strong winds. The pear tree will be ready sooner if the *du* stock is left taller; the stock may be surrounded by wickerwork filled with compacted soil to cover the top of the stock, then when it is windy the pear tree should be surrounded with woven bamboo screening to prevent any breakage.) Cut the bamboo obliquely to make a sharp slip and push it down between the bark and the wood to a depth of about 1 inch (2.5 cm). Cut a branch from the sunny side of a high-quality pear (a branch from the shady side will produce less fruit), about 5 or 6 inches (13–15 cm) long. Slice obliquely across its end, through the heartwood, so that it exactly matches the slip of bamboo in size. Use a knife to cut a very shallow ring around the pear branch just above the obliquely cut end and peel off the dark [outer] bark. (Do not damage the green [inner] bark, or the scion will die.) Pull out the slip of bamboo and push in the pear in its place, to as far as where the ring was cut, with the woody side against the wood and the bark next to the bark. When the grafting is finished, bind the top of the stock with silk floss [short, broken filaments from the silk industry] and seal it with clay. Earth up soil to cover it, leaving just the ends of the pear branches showing, and bank up soil all round. When watering, cover it with fresh soil after the water has soaked in; do not let the soil harden as it dries. In a hundred grafts, not one should fail. (Pear branches are very brittle, be careful when earthing up the soil not to knock them or they may break). [The striking similarity to the advice given by Fitzherbert (1548) is remarkable.]

Grafting by cutting a cross into the head of the *du* stock will result in failure more than nine times out of ten. (Doing so splits the wood and separates the bark, causing them to dry out.)

When the pear scions grow, if the stock produces leaves, they must be removed at once. (If not, the strength is divided and the pear must grow more slowly.)

When grafting pears, use side branches for those grown in orchards; use central branches for those planted in house courtyards. (When side branches are used, the trees are low so that picking is easy. Central branches produce upright trees that do not obstruct [the courtyard].) If basal shoots are used [as scions] then the tree will grow into an attractive shape, but will only set fruit after 5 years. If old lateral branches like pigeon's feet are used, they will set fruit after 3 years, but the tree will be an ugly shape.

Apple rootstocks

From the earliest days, apples and other fruits were grafted onto whatever rootstocks were available from hedgerow or forest, and rough empirical rules must have built up in communities as to compatibilities and the performance of wild material. Almost to the present day, rootstocks were gathered in this way in the remoter rural areas in China. By at least the 17th century in Britain (Markham 1640, Hartlib 1645, Austen 1657, Evelyn 1664, Rea 1665, Cook 1676, Worlidge 1676, Langford 1681a, b, "A Lover of Planting" 1685; also see Juniper and Juniper 2003), this random process was deemed unsatisfactory, and selections of so-called Paradise dwarfing rootstocks were used and vegetatively propagated. Vegetative propagation became a discipline in its own right (R. C. Knight et al. 1928).

The first written records of dwarf apple trees can be traced back to about 2,300 years ago to two students of Aristotle: Alexander, later the Great, and Theophrastus (Tukey 1964). Alexander, from his expeditions to the East, sent back a low-growing type of apple to the Lyceum in Athens. Theophrastus, who followed his mentor, Aristotle, as director of the Lyceum, recorded that the dwarf apple had probably long been grown in Asia Minor (Theophrastus 1916; also see Ferree and Carlson 1987). And the Romans were familiar with the growing of grafted dwarf trees of all types.

It has been suggested that the French Paradise apple was introduced from Armenia in the 17th century. Certainly by the mid-18th century the dwarfing of fruit trees of many kinds was well understood and described (Hitt 1755). Whatever the real source, by the second half of the 19th century many new dwarfing rootstocks had been introduced under the name; 14 kinds of Paradise are listed in Rivers's (1870) well-known text, *The Miniature Fruit Garden*.

The Armenian dwarfing rootstock known in its supposed place of origin as 'Marga Khndzor' appears to be identical, for all practical purposes, with 'French Paradise', now known as M8. The Georgian rootstock 'Khomanduli' appears to cover a range of clones individual genotypes, some similar to M8 and others closer to what is now known as M9. Other dwarfing rootstocks in Azerbaijan and adjacent southern Dagestan, 'Dipchek Alma', 'Kurl Almasy', and 'Yar Almasy', all similar to M8, seem to have been widely exported for use both in Russia and the West. However, trees grafted onto dwarfing rootstocks of this type were found not to be hardy in continental central Russia and suffered frost injury. Therefore, their use was restricted to the more temperate zone.

By the 15th century, mention is made of the dwarfing rootstock Paradise ('French Paradise') and the less dwarfing 'Doucin' ('English Paradise') (Tukey 1964). Swiss botanist and physician Jean Bauhin (1598) wrote extensively on grafting and describes his extremely dwarfing rootstock as a form of the Paradise rootstock. It is apparent, even by Tudor times, that "Paradise apple" implied not only something sweet but also a tree of dwarf habit (Rivers 1870).

Ruel (1536) referred to the paradisian apple and described it as though sent from heaven: very small, of honey-like sweetness, early as though spring-like. This little apple would appear to resemble the early, crimson, sweet, semi-feral apple of the Kursk and Voronezh regions of European Russia. Our word *paradise*, which now encompasses a spectrum of meanings, derives from the Old Persian word *pairidaeza* (*pairi*, around, and *diz*, to mold

The paradise garden. Drawing by Rosemary Wise.

or to form); the Greek word is *paradeisos,* meaning a walled garden. Regrettably, modern "paradise gardens" usually are constructed with corrugated iron and barbed wire (Plate 31) but still serve the original purpose of defending choice fruit and vegetables from marauders of all kinds, both two- and four-legged.

The earliest written Greek reference to "paradise" is probably from Xenophon (about 431 to about 352 B.C.), relating the story of Cyrus the Great's dedication to his gardens a century before. Longus, the 2nd or 3rd century Greek writer of the pastoral romance *Daphnis and Chloe,* described the *paradeisos* of Dionysophanes, which lay some 30 km (20 miles) from Mytilene on Lesbos, as lying on elevated ground and about 200 m (660 feet) long and more than 100 m (330 feet) wide: "It had every kind of tree—apple, myrtle, pear, pomegranate, fig and olive. On one side, it had a tall vine, which spreads over the apple and pear trees with its darkening grapes, as if it were competing with their fruit. These were the cultivated trees; and there were also cypresses, laurels, planes and pines . . . The fruit-bearing trees were on the inside, as though protected by the others. The other trees stood around them like a man-made wall, but these were in turn enclosed by a narrow fence" (Longus 1989).

In his famous *Paradisi in Sole Paradisus Terrestris* (*paradisi in sole,* park in sun, a pun on his own surname), John Parkinson (1629) shows, in a most elaborately engraved title page, Adam grafting an apple tree in an elegant paradise garden. Adam is modestly if not appropriately dressed for gardening activities and has a well-drawn 17th century face with fashionable moustache. Adam would appear to be grafting, unrealistically, a fully leafed-out apple tree, but we are entitled to allow the engraver a little artistic license. Parkinson was obviously firmly of the opinion that grafting was feature of a paradise garden.

So there is a strong hint here of the origin of grafting as a technique, and possibly its material, deriving from Persian gardens and Persian horticultural techniques. Intermediate stops on this progress west would seem to lie in Georgia and Armenia. With the benefit of hindsight, it can be seen that the poorly defined Paradise rootstock, having been secondarily selected perhaps in Armenia (where the word *pardez* refers to the garden around the house), almost certainly owes its origin to the great *Malus pumila* gene pool far to the east.

Today, in the fruit forest of the Tian Shan, it is possible to find among the cornucopia of fruit trees still surviving a diversity not only of fruit types but also of growth shapes: vigorously growing, markedly fastigiate, aggressively spinose, or striking dwarf forms. There is no difficulty in imagining that a dwarf form caught the attention of a would-be orchardist and was propagated vegetatively thereafter for grafting, regardless of the quality of its own fruit. The fruit from most currently used commercial rootstocks is small, tough, and almost inedible.

Title page of John Parkinson's *Paradisi in Sole Paradisus Terrestris*.

Nevertheless, in England it is frequently brought to celebrations of the autumnal Apple Days for identification. Such selection for rootstock characteristics was in parallel with, but independent of, selection for larger, sweeter, more colorful fruits for the table or striking blossoms for the ornamental garden (Chapter 7).

A difficulty encountered in attempting to propagate most apple cultivars by the usual vegetative means was slowly overcome by the development of stump splitting. In this method, a mature specimen of the chosen apple rootstock is cut almost to the ground, and the base of the stem is partially split using an ax and wedges. Then, over several seasons, radial sections of the base are split apart, again and again, to form still-rooted, if asymmetrically so, individual plants.

It would have been apparent to observant gardeners that Paradise was more compatible and longer lasting than the random rootstock collection from the hedgerow: "Many get *Crab stocks* out of the woods to graft upon for an orchard, but those kind of stocks are not (by far) so good, as such as come of *seede* or *kernells*, for many Reasons that might be shewed . . . . Plants coming of seede (and ordered as is shew'd) grow vigorously and seldome any fayle . . . besides they have an innate spirit (from the seede whereof they came) which makes them grow better" (Austen 1657; also see Meager 1670). This can be interpreted as evidence of the close relationship of the rootstocks to the scions, both from *Malus pumila* (hedgerow apples at the time were likely to be *M. sylvestris*, Plate 3), although their paths to the same nursery had been very different.

It is also generally true that passage through a seed stage tends to clean up plant material. Seedlings from sweet apples would, generally, be freer of endogenous diseases, which are now known to be caused mostly by viruses (though there are a few bacteria), than plants multiplied by vegetative propagation. The discovery of viruses lay far in the future, but the possibility of something inoculated into the sap, via budding or grafting, had been recorded as early as the 1710s by the Reverend John Laurence: "as of the Blood in the Body of an Animal . . . was able to be transferred across the graft union . . . . Effect of ting[e]ing by degrees all the Sap of a Tree" (Laurence 1716).

There was a widespread belief, current well into the 19th century, that regrafting was a process whereby ancient apple cultivars might be invigorated (Grindon 1885). There is no doubt that ancient, vegetatively propagated rootstocks might become debilitated by the slow accumulation of viruses, fungi, or mutations occurring in the tissue of the plant body, as foreshadowed by Laurence (1716). Thomas Knight (1797) was convinced of the degeneration, but there is no evidence that grafting would have reversed it. In fact, the opposite is likely to be true in that regrafted tissue will tend to spread any endophytic pathogens back into uninfected rootstocks. The only answer is modern, sophisticated tissue culture.

That by the early decades of the 20th century the then unconfirmed origin of the sweet apple in Inner and Central Asia was already apparent can be gleaned from H. M. Tydeman's (1937) observations on the suitability of Caucasian and Turkistani wild apple trees as potential rootstocks. (The name Tydeman is preserved in a number of well-respected and still widely grown apple cultivars raised under his direction at about this date.)

From 1912 onward, the East Malling Research Station in Kent, England, under the direction of Ronald, later Sir Ronald, Hatton, set out to test and characterize the various Paradise rootstocks of Europe, many of which, by apocryphy, had been used for centuries and in some cases possibly from Roman times. The confusion of names led to the abandonment of all the old names and the substitution, in the 71 collections Hatton gathered together, of Roman numerals I–XXIV for his selections, for example, Malling II for 'Doucin' and Malling IX for 'Jaune de Metz' (Hatton 1917). These dwarfing rootstocks were introduced at first as Malling (M) series in 1917, each clone with a numeral, for example, MI, MV, and MIX. In 1938 the prefix was changed to EM. Subsequent testing and introductions have raised the number of clones to 27, some of which are not dwarfing though M27 itself is markedly so.

When, by heat treatment of tissue-cultured material, it became possible to produce virus-free material, a joint effort between East Malling and the Long Ashton Research Station near Bristol, England, led to new rootstocks bearing the prefix EMLA; EMLA 9 is more vigorous than old M9. This characteristic should not surprise us since the rootstock has been parted from its debilitating viral symbionts, and it appears to be not just a virus-free derivative but a different clone. Such was the success of this program that it is estimated that 80% of apples throughout the world are grafted onto East Malling-derived rootstocks. Indeed, in a small experimental station in the home of the apple, Xinjiang Uygur, B.E.J. saw apples grafted onto a range of imported East Malling rootstocks.

A further collaboration between East Malling and the John Innes Horticultural Institute, then at Merton near London and now outside Norwich, England, was directed toward producing rootstocks with resistance to the woolly aphid (*Eriosoma lanigerum*). The aphid-resistant 'Northern Spy', which arose as a chance seedling about 1800 at East Bloomingfield, New York, was used as one parent, and the institute produced 'Merton 793', an invigorating rootstock with strong resistance to the attack of the aphid on the roots (Hearman 1936, Tobutt et al. 2000). From these hybridizations arose a range of aphid-resistant progeny. These East Malling–Merton hybrids, though very widely used throughout the world in the temperate and cool-temperate zones in the northern and southern hemispheres, have as a general rule not proved completely hardy under intensely cold and sustained continental winter conditions. In parts of Russia and China, rootstocks have been developed from the Siberian crab (*Malus baccata*, Plate 5) and hybrids therefrom with other northern species, for example, *M. orientalis* (now a synonym of *M. sylvestris*) and *M. pumila*. The whole of this technology of modern rootstocks is dealt with comprehensively by Rom and Carlson (1987) and Webster and Wertheim (2003), and DNA fingerprinting of rootstocks is discussed by Oraguzie et al. (2005).

CHAPTER 5

Westward migration of the apple

THE OLDEST RECORD of the sweet apple in China dates from 2,300–2,400 years ago. The Silk Roads, beginning as animal tracks, became the trading routes linking China with Rome and other Western centers. From Turkey, shipping routes took goods onward as far as the islands off western Europe. The Silk Roads passed around or through the foothills of the Tian Shan and, in summer, farther north through the range itself.

The Romans clearly distinguished between the imported sweet apple and the indigenous European apple, and it is possible they encountered already established sweet apples in Ireland, where those could have been introduced by the Celts in contact with the Berbers of the Mediterranean. All over western Europe there are Celtic place-names derived from their word for apple. The sweet apple soon entered the folklore, literature, and arts of the West, such that it is now probably more represented there than any other edible plant.

THE OLDEST USE OF TERMS generally accepted to refer to the cultivated apple seem to be found in early Chinese literature some 2,300–2,400 years ago. But it has to be accepted that, principally as a result of the depredations of the Cultural Revolution of 1966–1976, many ancient documents have been destroyed, and historical records of all sorts have disappeared. Complete libraries were burned during this destructive, iconoclastic period, and their knowledgeable staffs dispersed. Sadly, it may now never be possible to add much to Stephen Haw's researches discussed here.

Chinese names for apples

The earliest traceable occurrence of the word *nai,* normally translated as apple or sometimes crab apple, is in a fu, a prose poem, entitled *Shang Lin Fu* (on the

Upper Forest, an imperial park or garden) by Sima Xiangru (179–118 B.C.). According to Xin Shuzhi (b. 1893), the author of *Zhongguo Guoshu Shi Yanjiu*, studies on the history of fruit trees in China, the fu must have been written after the return to China of Zhang Qian, an emissary sent into central Asia by the emperor in 126 B.C., as it also mentions grapes, which are recorded as having been brought to China by Zhang Qian, though the source of that information may not be entirely reliable. If true, this would date the fu between 126 and 118 B.C. The fu is quoted in full in the biography of Sima Xiangru in the *Han Shu,* the official history of the Former or Western Han dynasty, a work compiled mainly by Ban Gu over a period of 30–40 years during the early part of the Later or Eastern Han dynasty. It was completed shortly after Ban Gu's death in A.D. 92.

There is also reference to *nai* in a book on the history of development of cultivated plants in China by Li Fan (1984), which seems to be very revealing. Li Fan mentions that "at the Jiangling Tombs have been discovered apple seeds, which very possibly are the seeds of Nai." There is an illustration, a rather poor black-and-white one, of the seeds with the caption, "Apple seeds from the Warring States period tombs at Jiangling in Hubei province." The Warring States period is usually considered to have been 475–221 B.C. Hubei (Map 2) is in eastern central China—the Chang, or Yangtze, flows through it—so it is not the first area that apples from the northwest would have reached. It is entirely plausible that sweet apples could have reached eastern China about 2,300 years ago. The trade routes into Inner and Central Asia were certainly operating by then, at least sporadically. The lack of mention in Chinese texts is not necessarily very significant, as very few texts survive from this early period. It is certainly very possible that apples could have reached eastern China two or three centuries before the earliest surviving mention of them, though it would be surprising, in Haw's opinion, if they were widely grown in China much before 2,400 years ago because all the common fruit trees, at least those of northern China, seem to be mentioned in the *Shi Jing,* an early compilation of poetry.

Chinese names for the apple have led to much confusion. The dictionary *Ci Hai,* sea of phrases, says that *nai yuan,* apple orchard, in ancient texts can also mean a Buddhist temple. This is explained in various ways, the simplest being that the White Horse Temple, the first Buddhist temple in China (as far as is known), had an orchard of apple trees. This is said not to be the most likely possibility, however. The dictionary quotes a fable: "Formerly in a kingdom in the Western Regions [now Xinjiang Uygur and adjacent areas] there was an apple tree that produced fruit and the fruit produced a girl. The king took her as a concubine. This girl established a Buddhist monastery on the site of the orchard, which was therefore known as 'Nai Yuan'." It is interesting that this story associates apples with both the remote western regions of China and with Buddhism.

Stephen Haw (pers. comm. 2000) has looked into the Chinese names used for the apple. Today, all are called *pingguo*. At one time it was thought likely that this name had come into use when apples from the West began to be cultivated in China during the 19th century, but this seems now not to be the case. The earliest traceable name believed to have been used for *Malus pumila* in China is *nai*. However, from an early period (at least as early as the 3rd century A.D., as it can be found in *Guang Zhi*, an almost lost work of reference of which only fragments and secondary sources survive).

Guang Zhi says (translations here and following by Haw), "There are three kinds of *Nai*: white, green [*qing*], and red. There are white *Nai* at Zhangye and red *Nai* at Jiuquan [both places in the Gansu Corridor]. In the western regions [including parts of what is now Xinjiang Uygur], there are many *Nai* everywhere. Households make *fu* of them [dry them] and store up to several tens or hundreds of bushels, just as jujubes and chestnuts are kept [in China proper]. During the time of the Emperor Ming of the Wei dynasty [reigned A.D. 227–239], the various princes were received at court and at night had bestowed upon them a container of winter-ripening *Nai*. Prince Si of Chen wrote in thanks, '*Nai* ripen in summer, but now they grow in winter. They are precious because they are out of season; gratitude is full though words are brief.' It was proclaimed that, 'These *Nai* came from Liang Zhou' [an administrative region roughly covering what is now Gansu and Ningxia Huizu with a small part of Qinghai and parts of eastern Nei Monggol (Inner Mongolia); during the Wei dynasty its capital was at Wuwei, northern Gansu]."

Register of Guests in the Palaces of Jin, another almost lost book about which very little is known, says, "In autumn, there are white *Nai*." *Miscellaneous Records of the Western Capital*, only a little earlier than *Qi Min Yao Shu*, says, "Purple *Nai*, green [*lü*] *Nai*. There are also plain *Nai* and vermilion *Nai*." *Guang Zhi*, which as explained is a lost reference work of which only secondary records survive, says, "*Liqin* [that is, *linqin*] are similar to red *Nai*." *Guang Ya*, a kind of dictionary dating from about A.D. 250, records, "*Nai* and *Linqin* are not grown from seed but are planted. (If seeds are sown, plants will grow, but the flavor [of their fruits] will not be good.) . . . Young plants can be produced in the same way as mulberry layers. (These fruit trees do not root freely, so cuttings rarely survive. Therefore layering must be used.) Another method is to dig a trench several feet from a tree, exposing the roots, from which shoots will then grow. Young plants can be produced similarly from any tree. Young plants should be planted as for peaches or plums. If trees of *Linqin* are beaten all over with the back of an ax during the first or second months [approximately equivalent to mid-February to mid-April], then they will fruit abundantly."

Nai was pronounced *nād* in the early Zhou period (about 2,100 to 1,900 years ago), and *nâi* in about A.D. 600 (Karlgren 1940). The latter pronunciation is is roughly that of the English word *nigh*. During the Han dynasty the pronunciation may perhaps have been similar to the first pronunciation given, with the final consonant pronounced very lightly or as a glottal stop. It may well be a loan word in Chinese—Could it derive from the same original as the Latin *malum* and the Turkic *alma* or *elma*? The occurrence as a major element of *ma* or *na* in all these words would be a strange coincidence if they were not all from the same original root.

Another name is also used: *pinbo*. The word is written with the same character for *pin* as used for *ping* in *pingguo* (Chinese characters usually have only one possible pronunciation, but some have two or three, or very occasionally four) or with a very similar character having the same pronunciation. The name is explained by texts from the 16th to 18th centuries as being a transcription from Sanskrit (used loosely to include Prakrit or Pali, or both—it tends particularly to refer to the original language of Buddhist texts). It is said to mean "correct and good."

Today, *pinbo* is used for a completely unrelated tree, *Sterculia nobilis* (Malvaceae), but that species occurs only in the extreme south of China, so it could not easily be confused with the apple trees of northern China. It is probable that in the absence of apples in the subtropical to tropical far south of China, the name, borrowed from the same source, was applied to a different tree. A similar transfer of names may have given rise in the West to the misconception that the apple was the tree in the Garden of Eden. When specifically used to refer to the apple fruit, the name sometimes has *guo*, fruit, added to it: *pinboguo*. The Chinese, however, tend to dislike words consisting of more than two characters. It is common for these to be abbreviated, so the middle character, *bo*, disappears, leaving *pinguo*. The final *n* of *pin* may then have become *ng* because of the following *g* in *guo*, giving *pingguo*.

The first occurrence of *pingguo* seems to be in a work called *Qun Fang Pu*, record of all sweet flowers, compiled by Wang Xiangjin about A.D. 1621, certainly before 1630:

Nai are also called Pinbo or Pingguo. Those from Yan and Zhao [roughly equivalent to modern Hebei and eastern Shaanxi provinces] in the north are very good. They are grafted onto Linqin stocks [possibly *Malus asiatica*].

The trees are bushy and erect, with green leaves like those of Linqin but bigger. The fruits are like pears but round and smooth, green when unripe, ripening to half red and half white, or entirely red, shiny and delightful, with a spreading fragrance and a sweet taste. When not yet ripe they have a texture like cotton wool, when too ripe they become soft and gritty and unfit to eat. They are at their best when eight- or nine-tenths ripe.

This seems to be reasonably sound evidence for the synonymy of *nai* and *pingguo;* other texts from about the same period regard the two as the same. *Guang Qun Fang Pu,* enlarged record of all sweet flowers, of about 1708 says, "Pingguo: the herbals do not mention Pingguo, but instead use 'Nai'. Another name is 'Pinbo'."

In modern China a distinction is made between cultivars of apples called *mian pingguo* and other ones. This *mian* means "silk floss" and may refer to the texture of the unripe fruit; it is also used with the derived meaning "soft," which might refer to the texture of the fruit when too ripe. Most modern Chinese writers on apples seem to consider these *mian pingguo* as old native cultivars distinct from those imported from the West after about 1860. It seems that the Western cultivars generally keep much better and do not readily become overripe, as the old Chinese ones do. This is probably the main reason why the old Chinese cultivars seem largely to have been displaced by phase 1 Western ones.

Linqin is an odd word that does not look Chinese. It seems that it was written in several different ways in early times, during the first few centuries A.D. Before A.D. 500, forms such as *liqin* and *laiqin* (with different characters for writing *qin*) were often used, but since then *linqin* has been a fairly stable form that has continued in use to the present. It seems highly probable that a non-Chinese word was being transcribed into Chinese, but from what original language is very uncertain.

Stephen Haw concludes that the early records, fragmentary as they were and then further degraded in the Maoist period, were mostly compilations from secondary or even tertiary sources. For example, Jia Sixie's *Qi Min Yao Shu* is really a collection of quotes from earlier texts relating to plants that were presumably not familiar to the author, many about plants from what is now southern China. China was divided into more than one state during Jia's lifetime (and for many years before and since).

Haw considers that some notes on, and speculation about, names for the apple may be relevant here. There is little doubt that *nai* and *linqin* were definitely kinds of *Malus*, but it is now impossible to be absolutely certain what species were meant. Later usage strongly suggests that *nai* is *M. pumila* and that *linqin* is *M. asiatica.* There are not really many other possibilities—most other apples occurring in or grown in China have very small fruits that are scarcely worth eating and would not really need cutting up before being dried.

The Tokharian, or Tocharian, word for apple is unknown, though the people so called almost certainly grew the fruit. In what is now Xinjiang Uygur and adjacent regions during the final millennium before the present era, several very obscure languages belonging to at least three major language groups were probably spoken (J. M. Renfrew 1987). Tokharian was Indo-European: there were probably other related languages spoken in the area of which we now have virtually no knowledge. Some peoples in the region probably used proto-Mongol languages. The language of the Huns (known to the Chinese as Xiongnu) is very little known but may have been Paleo-Siberian.

Silk Roads

Die Seidenstrassen, the Silk Roads, was a term first used by the coiner of the word *loess* for the windblown dust soils of much of China, the German geologist, traveler, and writer Ferdinand Paul Wilhelm von Richthofen (1833–1905), the uncle of the Red Baron, Manfred von Richthofen, the fighter pilot of World War I fame. The term probably first appeared in print in 1877. It is regrettable that most of Richthofen's valuable work concerning the region (Richthofen 1883) has yet to be translated.

Fundamentally misnamed, but a label with ineradicable romantic connotations, the Silk Roads were of enormous economic importance and definitely used thousands of years before silk was invented; they reach far back to the dawn of mankind. They began, almost certainly, as animal migration tracks, connecting sources of water and grazing, long before the emergence of humans and even longer before the establishment of agriculture about 13,000 years ago. But what we now understand as the Silk Roads were the ancient trade routes linking China, principally under the Han and T'ang dynasties, with imperial Rome and other major centers in the West. As noted in Chapter 3, the Chinese do not use the term Silk Road but Horse Road.

The main route was about 10,000 km (about 6,000 miles) long, stretching west from the ancient Chinese capital of Ch'ang-an (now Xi'an, Shaanxi; Map 6); Beijing was then only a small urban area and did not become the capital until much later). The main route ran through the fertile Gansu Corridor, across the dry and dusty Gobi, and around or through the lower passes of the Tian Shan. The track then passed through the ancient, almost mythical kingdoms of Sogdiana and Bactria, the land of the Calmucks and Parthians, through ancient Media and Persia (Map 8), skirting the southern edges of the Caspian and Black Seas, through Mesopotamia at Babylon, and to Damascus, Memphis, and Alexandria, or to Antioch, or through Cappadocia to Constantinople and other ports of what is now western Turkey. Lesser land routes ran west of the Caspian Sea and north of the Black Sea through Colchis and Scythia and on into the Danube Valley. At the same time, the sea-lanes from the eastern Mediterranean may have spread trade goods and ideas throughout the Mediterranean world and beyond to the whole seaboard of western Europe (Cunliffe 2001).

The Silk Roads, as such in the narrow sense, began to be used about 2,100 years ago, after the Han dynasty took control of large areas of Central Asia, and were in vigorous use for the next 500 years or so, at the height of Rome's financial, military, and political power. Goods varied. For example, it has commonly been assumed that pottery (from a very early date, and from about 1325 the famous blue and white porcelain), unlike silk or similar high-value goods, was too fragile to be carried on camel or horseback. Complete junk cargoes, salvaged from the depths of the China Sea, seem to suggest that shipping was the preferred mode of transport. There is good evidence too that cargoes of pottery,

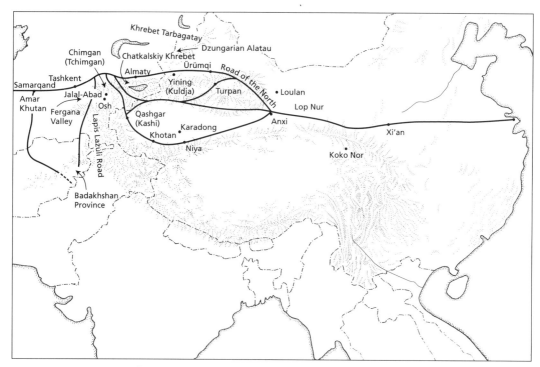

MAP 6. Eastern section of the trade routes. Also shown are some of the principal and still extant areas of fruit forest, for example, Khrebet Tarbagatay.

including blue and white, were transported by the Chinese as far as the Red Sea in the 14th century (Carswell 2000a). But there is also evidence that small, land-carried loads of pottery and porcelain were packed into cases that were then filled with a mixture of sand, earth, soybeans, and wheat (Carswell 2000b). The cases were dampened, and the matrix set rock hard. At journey's end the cases were washed, dissolving the cement, and the contents retrieved.

The shortest routes from east to west skirted the deserts. But when the summer's heat grew too oppressive, routes on higher land were used. From the Gansu Corridor the track ran across the Gaxun Nur, or Gashiun Nor, salt basin and into the mountains of the extreme eastern end of the Tian Shan—what the Uygur call the Road of the North. This northern road was longer, more mountainous, and made the horse-based and yurt-dwelling caravanners somewhat more open to attack by bears (Plate 24). But in summer and autumn, until the snow and ice closed the high passes in early November, they passed through vast stretches of intact fruit forest interspersed with high, grazing meadows. The Road of the North would, from time to time, probably have been used by the Turpan and Tokharian people about 4,000 years ago. It was certainly used by Genghis Khan and his army in in the early 1200s.

The Roads, then, almost certainly began as animal migration tracks as great aboriginal herds of wild horse, ass, donkey, and camel ranged the steppe in search of fresh grazing, oases in the summer's heat, and vegetation in the winter's cold. Only at a much later date were these tracks adopted by humans as convenient routes of commerce. The grazing animals of the steppe land would also have pioneered such tracks as the Road of the North when the supply of the steppe's grasses and herbs withered at the end of long, hot, dry summers. For ancient ranging quadrupeds, and later for humans traveling with animals, the routes linked oases that guaranteed sources of fresh water and also grazing and browsing. Two wetter climatic periods about 3,000 and 8,000 years ago would have encouraged and sustained both the domestication of the horse and major movements on these trade routes (Broecker and Liu 2001).

It is a strange fact that over almost the entire length of the Silk Roads there is practically no significant supply of building stone, nor of coal with which to make fired brick. Houses and other structures, even palaces and city walls, have always been built of the readily available, low-energy, low-technology, but effective mud brick. Taller structures combined timber studding and mud-brick infill. But without careful maintenance, thatched roofs disintegrate, walls dissolve, and finally the whole structure reverts to arable soil. Timbers are recycled or converted to firewood. Finding any archaeological evidence necessitates digging.

Lapis Lazuli Road

Possibly far older than the main east–west trade routes as strictly commercial routes, and perhaps dating back to some 7,000 years, is the Lapis Lazuli Road (Map 6). The precious mineral, its name a combination of Latin *lapis,* stone, and Arabic *azul,* blue, formed the intense blue coloring of the ancient world and is used to the present day as both jewel and pigment. King Tutankhamen's mask was made, in part, of lapis lazuli, as was some of Queen Pu-abi's treasure at Ur. Lapis lazuli was first, and still is, mined at Sar-i-Sang in Badakhshan province of Afghanistan some 250 km (160 miles) north-northeast of Kabul. It would have been transported north, principally. First it may have been carried on the backs of porters, but when the horse and the camel were domesticated, progress would have been much more rapid. The obvious route through what are now the cities of Osh and Jalal-Abad of Kyrgyzstan would have taken supplies up to the main east–west routes. Then it would have been transported west to the great empires of Babylon, Persia, and Egypt, and later east to imperial China. Lapis lazuli traveled far. In its own way it cannot have failed to expedite the movement of other substances and crops, including fruits, as well as ideas and techniques, along the great east–west routes.

Whatever the routes, and whatever the cargo, the caravans across the deserts of Inner and Central Asia would often travel at night to avoid the blistering heat of the day. Such caravans would traverse stages from one water source to another, of distances rarely greater than 32 km (20 miles) in 10 hours of travel (Cable and French 1936, 1942). The routes continued in major commercial use until the 6th or 7th century A.D. By comparison with modern high-speed transport of goods, it should be remembered that any consignment, such as bales of silk, would have taken at least a year to reach its destination. But there is excellent archaeological evidence that manufactured goods did travel the whole route from Ch'ang-an to Roman Londinium (London, England).

Beginning with the earliest attempts at cartography in the 16th and 17th centuries, the existence of these ancient trade routes was long unknown or at least not depicted, and what we think of now as great cities along these routes— Tashkent, Shymkent, Bishkek, and Almaty—were just small settlements of yurts or not recorded. Archaeological evidence of a clustering of yurts does not last long. To the south, the ancient Buddhist cities of Khotan (Hotan), Yarkand

(Shache), and Kashgar (Kashi) seem to have been somewhat more robust. Near the ancient roads can still be seen what are evidently artificially leveled platforms where an encampment of yurts might have been assembled, and to the back of these platforms often lies a partial ring of apple and other fruit trees, the living results of the ancient debris of many a human and animal meal (B.E.J. pers. observations 1999).

The roads flourished under the T'ang dynasty (A.D. 618–907), but as this civilization crumbled, so did the way of life along the Silk Roads: way stations, forts, monasteries, temples, towns, and even complete cities vanished almost completely, particularly since they were predominantly built of mud brick and were not rediscovered until the end of the 19th century (Hopkirk 1980).

There were many reasons for the slow but inexorable death of these long transcontinental routes, but the principal one was the progressive drying of Inner and Central Asia. The falling level of the Caspian Sea and the intermittent flow of rivers point to a long and sustained dry period, to today (Lamb 1995). The whole area has been slowly drying ever since the end of the last glaciation some 12,000 years ago, with occasional improvements, but there seems to have been an acceleration in this widespread desiccation shortly after the Roman imperial period. Glacier-fed streams that flowed from the Tian Shan and the Pamirs slowed and in some cases petered out. Irrigation systems, often of great complexity, were built but mostly eventually failed to sustain urban environments.

From the earliest times, probably particularly as the natural desiccation of the region began to tighten its grip, local agricultural communities developed sophisticated underground canals. These vein-like systems led from the glacier-fed slopes of the mountains on both sides of the Tian Shan and parallel ranges, often far across but under the desert, to oases. Such technological skill saved precious water supplies and maintained agriculture but profoundly modified the foothill vegetation. Some of this technically ingenious venous system was monumental in scale. These channels, called qanats in ancient Persia and kuvur in Uzbek, often ran for great distances; for example, one more than 70 km (43 miles) long is known from Persia. They were often at great depths, some as much as 300 m (980 feet) beneath the surface. Given the advantage of a generous, high-quality water supply from the mountains and the intense summer heat and sunshine, very early crops were, and still are, possible. Such a bonus may have given rise to the Turkish word *turfanda,* referring to very early season exotic fruits, which may derive from the faraway southern Taklamakan city of Turpan.

In the end, the proselytizing warriors of Islam from the west overwhelmed the struggling survivors. The mud city of Niya, now an almost unrecognizable ruin on the southern edge of the route around the Taklamakan, died in this way at the end of the 3rd century A.D. when the Chinese temporarily lost control of the Silk Roads. Niya withered away so that it disappeared completely from maps and was not rediscovered until the turn of the 19th and 20 centuries. The final blow to Chinese control of Central Asia came when a Muslim army defeated a Chinese army at Atlakh near the Talas River in what is now Kyrgyzstan in July 751—the battle lasted 5 days. This victory ensured that the predominant religion of Central Asia became Islam, as it remains. The victory also set in motion the collapse of the T'ang dynasty.

Under the Ming dynasty (1368–1644), the Silk Roads were finally abandoned and China shut herself off from almost all contact with the West. This abscission led to further isolation of the Tian Shan. However, Chinese sea traders, often with full cargoes of porcelain, were still trading throughout the China Sea and probably reached as far as the eastern coast of Africa and the Red Sea. In 1405, the Yongle emperor promoted a series of seven maritime missions to the West, and his first convoy comprised no fewer than 317 ships. But shortly after, China effectively closed her borders, by all routes, to foreign influence.

As a trade route, the fate of the Silk Roads was finally sealed when first the Arabs, then the Portuguese, followed by the Dutch and then the British, established trade routes round the Cape of Good Hope, through the Gulf of Arabia, round India, and to Cathay. Sea travel became faster, cheaper, and more reliable, in fact the only practical way for large cargoes to travel. The function, if not the myth, of the Silk Roads sank into oblivion. So deep was this oblivion that on European maps of the 16th to 19th centuries, although the Pamirs, for example, are marked, there is no reference at all to the by then well-established great cities of the eastern section of the Silk Roads, and no indication of the route itself. However, it is interesting to note that what is certainly the Lapis Lazuli Road from Sar-i-Sang up to Osh is usually depicted. Lapis lazuli was and still is in heavy demand. That route, although now busier than it has probably ever been with other valuable if perhaps less agreeable products of Afghanistan, was and still is the only way for lapis lazuli to reach the West.

It has been argued that these overland routes were of less significance than sea routes (Ball 1998). There is no doubt that there were some sea routes, from very early times (Cunliffe 2001). A very heavily laden merchant ship, about 40 m (130 feet) long, has been discovered near Varna, ancient Odessus, off what is now the coast of Bulgaria. It apparently came from the port of Sinope (now Sinop on what is now the Black Sea shore of Turkey). It appears to have taken its counter-clockwise journey around the Black Sea sometime between 2,486 and 2,276 years ago, very close to the time of Alexander the Great. It was laden with amphorae principally containing dried catfish steaks and olives. Xerxes saw merchant ships carrying wheat from the Ukraine to the Peloponnese. The sea route indicated by these observations was from the Crimean Peninsula through the Bosporus and Dardanelles to the Aegean. It would have been then just a short sea journey to Ostia at the mouth of the Tiber, downriver from Rome.

Direct archaeological evidence for the existence of the land routes comes from at least two sources. First, there are the finds made in an excavation in the Spitalfields area of central London. In 1999 a stone sarcophagus, containing an elaborate lead coffin, was found there, just outside the walls of Roman Londinium. In it was the body of an early 4th century woman who died, without major trauma, in her early 20s and whose place of birth was probably not Britain but southwestern Europe, possibly Spain. Her very sophisticated grave goods of jet and glass indicate that she came from a wealthy family. All of this assemblage is now on permanent display in the Museum of London. Her dress is particularly intriguing: a full-length garment of silk, partly gold-embroidered. The silk was produced in China, as identified by the dimensions of the filament, and the raw silk had been woven in what is now Syria. It is not credible, except possibly for the last stage from the Roman port of Ostia to Londinium, that sea transport played any part whatsoever.

Second, excavating at Loulan on the northern edge of the Lop Nur basin in the eastern Taklamakan in 1906–1907, Mark Aurel Stein found a small, intact bale of yellow silk among the ruins (Walker 1998). Loulan was occupied by troops of the Chinese imperial army as one of their most western outposts, in the first three centuries A.D. It can only be surmised that this postal item, like many since, failed to reach its destination in the West.

Apples in Europe

It can be deduced from present-day distributions, thin as they are, that a few species of apples, mostly of quite limited geographical distribution, were present in eastern, central, or western Europe immediately after the receding of the glaciers: *Malus dasyphylla*, *M. sylvestris* (Plate 3), which includes *M. orientalis*, and *M. florentina* and *M. trilobata*. But how common were they? There is one small piece of evidence. In an exhaustive treatment of the trees and timbers of the ancient Mediterranean world, covering an area from east of the Tigris in Mesopotamia, south to Memphis on the Nile, and north beyond the Alps (Meiggs 1982), there is no mention at all of any species of apple. A range of timber species, their sources, local or imported, and their uses, from cabinetwork to bridge building, are examined. Ebony came from India, walnut from Asia, principally Kyrgyzstan, but there is no mention of any apple as a timber tree.

Yet there are numerous mentions of the sweet apple in the discourses of Roman horticulturists such as Pliny and Columella. An interpretation is that the wild apple species were rare, even then, and of no significance as either ornamental or structural timber. Tacitus (emperor A.D. 275–276), probably born somewhere on the Danube, noted that the Germans ate wild apples, *agrestia poma*, whereas the Romans preferred cultivated apples, which they called *urbaniores*. Here the distinction between *Malus sylvestris* (*agrestia poma*) and *M. pumila* (*urbaniores*) was abundantly, one might say snobbishly, clear. To show the esteem in which *urbaniores* were held in Rome, glass bowls holding apples with other fruits are depicted in murals (Wasserman et al. 1990), and on the walls of rooms of the House of Livia in Pompeii, a complete apple orchard is beautifully rendered. There can be no doubt that these apples are *M. pumila* and not a native crab.

The orthodox view, and one no doubt that the Latins would have encouraged, is that the Romans brought the complete technology, learned from the Greeks, of apple grafting, cultivation, harvesting, storage, accumulation of new varieties, and probably cider making to northwestern Europe (French 1982). "The apple was, in all probability, introduced into Britain by the Romans, as well as the pear; and like that fruit, perhaps reintroduced by the heads of religious houses on their establishment, after the introduction of Christianity" (Loudon 1844).

It is generally assumed that the Romans came to Britain in search of copper and tin, and of an abundant supply of high-quality wheat with which, through their port at Ostia, to feed the potentially riotous population of Rome. No doubt they found a sophisticated Celtic cereal agriculture with a far higher quality of wheat than could be grown in Rome. But did they find anything else? Despite Loudon's assertion, is it possible that the Romans also found, but did not acknowledge, rare, outstanding, seed-grown, own-rooted individuals of the sweet apple? Is it conceivable, too, that the large, possibly 3,000-year-old apple discovered at Navan Fort, County Armagh, Northern Ireland, might be an early Celtic-Phoenician seedling import of *Malus pumila* and not an extraordinary example of the native bitter crab, *M. sylvestris* (Kennedy 1997)? The Romans were acknowledged masters in denigrating the achievements of their subject peoples (except for their war-like qualities, which they needed to subdue), and there are no written records of Celtic fruit growing in Britain.

Archaeological evidence of the Celtic peoples is widespread, and intriguingly, their grave goods of gold and pottery indicate powerful influences from Greece. Map 7 shows the distribution of Celtic archaeological relics about 2,400 years old. At that time not only were they building sophisticated stone dwellings in the Shetland Islands off northern Scotland, but in their eastern reaches they

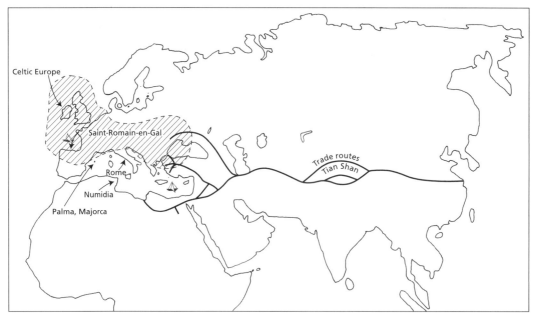

MAP 7. The extent of Celtic influence about 2,400 years ago, overlapping at the eastern end with the east–west trade routes.

must have been in contact with the highly advanced agriculture and horticulture on the edges of the Persian empire. What is more, as is well attested by their grave goods and by the Roman accounts of battle, the Celtic peoples were familiar with, and widespread users in all modes, of the horse.

From the beginnings of agriculture in western Europe, say 6,000 years ago, there seem to have been excellent sea-trading routes around the whole of the western maritime fringe as well as those circumnavigating the Black Sea. Around 3,400 years ago, a small trading vessel, which seems to have been a tramp steamer of its day, taking cargo from port to port opportunistically rather than on a predetermined itinerary, sank off the southern coast of what is now Turkey at present-day Ulu Burun, between Kalkan and Kas (Map 8; Cunliffe 2001). It was carrying about 6,000 kg (13,000 pounds) of copper ingots, probably from Cyprus, ingots of tin, possibly from Etruria or even as far as Galicia or Cornwall, glass from Syria, ivory from Africa, and amber from the Baltic (Map 9). The evidence of an important Bronze Age international sea trade in exotic, imperishable materials is irrefutable. What perishable goods, including food supplies, might this ship also have been carrying? Might apple seeds, along with seeds of other crops, have been carried to western Europe? It is believed that grafting was well established by about 2,400 years ago, and within a few years Alexander the Great's crews

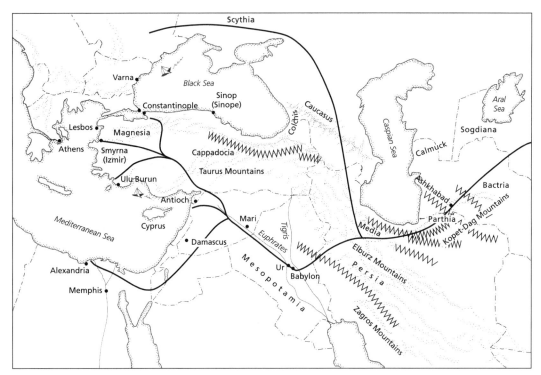

MAP 8. Western section of the trade routes, with fault lines shown as zigzag lines. Bronze Age sea routes are indicated by sailing vessels.

practiced sea battles using apples as projectiles. These later dates fall within the range of the Mediterranean sea peoples. Apples, if they were part of the onboard food supplies, might have been derived from either grafted or elite random seedling *Malus pumila* trees—it would make no difference in the spread of the fruit.

Celtic place-names and the apple

All over western Europe, and vestigially in Britain, there are pre-Roman, probably Celtic place-names that suggest the cultivation of apples: Aveluy in Picardie, Haveluy also in northern France, Availles in Ille-et-Villaine and in Poitou, Evaillé in Pays de la Loire, Havelu in central France, Avallon in Bourgogne, and Vallon-Pont-d'Arc in Ardèche. In addition there is Avilar in northern Spain, which likewise came under the influence of the Celtic peoples. In Britain there is only the hazy recollection of the Isle of Avalon, or Ynys Avallach. The tangle of myths and unsubstantiated claims surrounding the existence of the so-called Isle of Avalon locate it somewhere from Brittany to Edinburgh. But it seems likely that embedded within these legends and fragmentary archaeology is a favored place where choice apples grew, and that these were sweet apples and not the miserable, bitter crabs of the forest (R. Palmer 1996).

All these apple place-names carry the root *av* and not the supposed Saxon *apf* or *äpp* of the Saxon takeover. The Celts may simply have been recording places where the little wild apple, *Malus sylvestris*, grew. They would have been well acquainted with that in their passage from the Middle East through eastern to western Europe. But it is also legitimate to speculate that they brought with them seeds, at least, of the sweet apple, if not the skills of grafting. From these seeds grew many indifferent specimens but perhaps also a few spectacular, elite apple trees with renowned fruit, which gave their name to a particular locality. Such a speculation might also explain the otherwise puzzling references to outstanding apples in Ireland that lay beyond the Roman writ but within the range of Celtic migrants. This interpretation may also explain occasional references, often dismissed by later authorities, such as "the apple Whitaker (1805) conjectures to have been brought into Britain by the first colonies of the natives" (Loudon 1844).

Romans in Britain

Although there are excellent written records and depictions of Roman apple culture in mainland Europe, there seems to be no record of their transfer to Britain after the Roman conquest. The Romans were well acquainted with the whole range of techniques of grafting. It must be assumed that they came to Britain in A.D. 43, tasted the local apples, and said something upon the lines of *Vobis malum sit*, evil be unto you (the pun intended as a reinforcement of the Latin confusion of *mal* words; Chapter 3). Perhaps they found both the wild crab (*Malus sylvestris*), and here and there, rare but treasured, seed-derived examples of the Inner and Central Asian sweet apple (*M. pumila*). These tiny pockets of esteemed sweet apple would have been commemorated in place-names in northwestern Europe incorporating the Celtic *av*.

It seems likely that the Romans found the local apple, *Malus sylvestris*, in common use, particularly in its dried form, but despising this local example of *agrestia poma* chose to graft their sweet scions onto the decapitated stumps. There is indirect evidence of this practice at least in Germany in the Roman period (Hehn 1902). Roman craftsmen would have been well aware that the upper parts of the ancient *M. sylvestris* trees were invaluable as iron-hard timber for the teeth and cogwheels of the watermills and windmills that they introduced to grind wheat for export to Rome. Such decapitations could help explain the extreme rarity of *M. sylvestris* in lowland Britain. It has to be admitted, though, that neither by written testament nor by the footprint of a Roman place-name, nor by the image in any mosaic pavement, is there any direct evidence of apple cultivation over the nearly 500 years of Roman influence in the British Isles.

The empire of Charlemagne

By the end of his reign, Charlemagne (A.D. 742–814) effectively ruled almost the whole of France to the Pyrenees, much of what is now Germany and Austria, and most of Italy. Charlemagne caused to have drawn up about the year 800 a list of some of the fruits grown in his kingdom: "De arboribus volumes quod habeant pomarios diversi generis, . . . Malorum nomina: 'Gozmaringa': 'Geroldinga': 'Crevedella': 'Spirania': 'Dulcia': 'Acriores' omnia servatoria; et subito comessura; primitiva. Perariciis servatoria trium et quartum genus, dulciores et cocciores et serotina," Concerning trees it is our wish that they have fruit trees of different types, . . . The names of the apples are 'Gozmaringa', 'Geroldinga', 'Crevedella', 'Spirania', 'Dulcia', 'Acriores'; all are for preserving, or for eating straight away, early flowering. Perariciis [cannot be translated] preserving apples of the third and fourth type, sweeter and redder and late flowering (Fois 1981, Harvey 1981, 1992). Gozmaringa (now Gomaringen) is a little town in Baden-Württemberg; Gerold-inga may be Goldingen in the district of Courlande in present-day Lithuania. Crevedella has not been identified. *Spirania* implies fragrance, *dulcia* sweetness, and *acriores* great acidity and the ability to last a long time (Leroy 1873). The details in these names imply grafting; some of the apples may have arisen from chance seedlings in local districts. Sadly, none of these cultivars seems to have persisted.

Berbers and Moors

There is yet another strand to add to the complex web that connects the story of the spread of the sweet apple to the West. Long before Rome, and even longer before the emergence and spread of Islam—Muhammad lived from about 570 to 632—sophisticated peoples, including the Berbers, had spread along the North African coast. They had established a civilization that extended from Egypt west to the Atlantic and crossed over into what is now southern Spain and Portugal. The origins of the Berbers are obscure, but in some of their architecture and art forms they appear to have had contacts with early Greek civilizations. At times, the Berber influence was so strong that in their ancient lands of Numidia, in present-day Algeria, the Romans in the 2nd century feared that another Carthage was emerging and moved to suppress the development. There can be little doubt that for hundreds of years in what is now southern Spain and Portugal, the

Berbers must have come up against, and no doubt communicated with, the Celtic peoples of mainland Europe.

The rule and culture of Islam was imposed from Damascus and Baghdad along the North African coast, through to Toledo in Spain, from about the year 670 to the defeat in Spain in 1492. Charlemagne was in regular contact, sometimes of a hostile nature, with these Muslim peoples, as recorded in *The Song of Roland*. There is growing evidence that a wide range of fruits and vegetables, apart from the well-attested examples of some citrus fruits and the carrot, entered western Europe at various times along this corridor. There were periods when the Islamic Arab empire reached from the Tian Shan and the Indus in the east deep into Europe and all the way along the African coast to southern Spain and Portugal, and the ships of this empire could carry perishable goods with speed and efficiency. This religious dominance was not imposed in a vacuum but upon existing civilizations, sometimes treating them, as in the case of the Berbers, as subject peoples, but it was often characterized by sophistication in horticultural practice. It seems very likely that apart from the more exotic fruits of the subtropical world, such as the apricot (which word itself is Arabic), citrus, and olive, some apple cultivars with a low cold-chill requirement were brought this way into Spain and Portugal.

In 1080 the Moors, who then controlled Toledo, may have commissioned Ibn Bassāl to write *The Book of Agriculture* (Harvey 1992). This book contains information of many crops, including apples. Its translation into Castilian was probably made at the instigation of Alfonso X, the Learned, king of Castile 1252–1284, whose half-sister, Eleanor of Castile (1246–1290), took a copy of the book to England with her when she went to marry Edward I, king of England 1272–1307, at the age of 10 or 11. With the help of the text, plus imported craftsmen and plant material, all testifying to the poor state of horticulture and plant knowledge in Britain at the time, she established a very modern garden at King's Langley in Hertfordshire. Following this brief dawn, however, agriculture and horticulture deteriorated in Britain and did not begin to revive to anything approaching the Continental level until the end of the Tudor period.

Post-Roman place-names and the apple

As the power of Rome progressively weakened during the 5th and 6th centuries, Germanic tribes began to press into western Europe. According to the Roman emperor Tacitus, these groups occupied the area of modern Holstein in Germany about A.D. 100. Ptolemy had them in the same place in the mid-2nd century, but between the years 250 and 450 these Saxon peoples began to move west, first settling among the Frisians on their islands. Various of these Germanic tribes, now comprising Angles, Franks, Saxons, and Jutes, continued to move west. The Franks predominantly came to occupy what is now northern France. Some of the other Germanic tribes were, at first, invited to post-Roman and Celtic kingdoms in Britain as mercenaries to fend off attacks from the Picts and the Scots to the north. In a very complicated situation, the original Romano-British occupants were slowly overwhelmed by the invaders from the east, but the takeover appears not always to have been hostile and was sometimes by mutual agreement. On the other side of the Channel, the Franks, of whom some small populations may have established themselves in Britain, consolidated their hold and, under Clovis, king of the Salian Franks 481–511, effectively united much of modern-day France.

But there were striking differences in the subsequent histories of the regions as these Germanic tribes displaced, murdered, miscegenated, or reached agreements with the local post-Roman inhabitants. On the whole, the Roman estate and farm names in northern France were retained. Apple names in France are predominantly Celtic, though the French revere the apple as much as, if not more than, the English. French production of apples in all its forms always has been far larger than that in Britain. But the old Roman names virtually disappeared from Britain, and the only nomenclatural relics of 500 years of occupation and influence are a handful of town and city names incorporating *cester* (for example, Cirencester in Gloucestershire, and Portchester in Hampshire), *foss, pons,* and *porta.* In fact, there are far more relics of Britain's Celtic past than that of the Roman period; the name of almost every river, stream, forest, and hill can be ascribed to the pre-Roman inhabitants. A further and equally odd distinction is that over almost all of mainland Britain, along with the very small but recognizable Frankish regions, the Angle-Jute-Saxon invaders left behind, usually as villages or small towns, the name of the apple (usually as a prefix), for example (Ekwall 1991, R. Palmer 1996, etc.):

Apley and Apperley (apple-tree wood): Gloucestershire, Isle of Wight,
    Northumberland, Shropshire, Somerset, West Riding of Yorkshire
Apperknowle (apple-tree hill): Derbyshire
Appleby: Lincolnshire, Westmorland
Appleby Castle: Cumbria
Appleby Magna and Parva (apple village or homestead): Leicestershire
Appledore (apple tree): Devon, Kent
Appledram and Applesham (*ham,* a village, estate, manor, or homestead):
    Sussex
Appleford (the ford by the apple trees): Berkshire, Isle of Wight

Applegarth (apple orchard): North Riding of Yorkshire

Appleshaw (apple wood): Hampshire

Applethwaite (a clearing where apples grow): Cumberland, Kent, Lancashire, Norfolk, North Riding of Yorkshire

Appleton (where apples grow, or an orchard): Berkshire, Cheshire, Cumberland, Kent, Lancashire, North and West Ridings of Yorkshire

Appletree: Northamptonshire

Appletreewick (*wic*, a dwelling, village, street, or farm): North Riding of Yorkshire

Appley Bridge: Lancashire

Appuldurcombe (apple-tree valley): Isle of Wight

Are these names relics of *Malus sylvestris* or of *M. pumila*? *Malus sylvestris* was usually an indicator of a boundary, and the sweet apple, *M. pumila*, would have been nearer settlements (Oliver Rackham pers. comm. 2005). Since virtually all the suffixes in these place-names are of Anglo-Saxon or less frequently Danish (*by* is Old Danish for farm) origin, it must be assumed that on the whole they represent either the establishment of new settlements or a takeover of existing old Roman orchards by the Anglo-Saxon invaders. Either way, it is very unlikely that these names, giving as they do a record of fruit of distinction, refer to the little, local crab apple, *M. sylvestris*. The Saxons and Danes would have been well aware of that bitter little crab as they migrated through northwestern Europe and could not have held it in any esteem. It seems far more likely that the place-names represent relics of orchards of *M. pumila*, imported and grafted by the Roman colonists, or even earlier, and treasured by the later owners. The orchards may have tumbled down to mixed seedlings, many of indifferent quality, but among which by chance were large, sweet, elite genotypes of distinction. The diversity of apple form, texture, and sweetness was almost infinite, the potential to select and propagate a local apple of quality an added bonus. But the clear dividing line in place-names provided by the Channel remains a mystery, though the Franks were closely connected in every way with the Saxons. It is interesting, too, that pear names are rarer than apple names in Britain. Those may derive from the Old English *pirige* and are found in names like Parbold, Parham, Parley, Preshaw, and Prested.

The Battle of Hastings

If, and this is speculative, the history of the Saxon period records the large, sweet apple, as seems reasonable, and not the local crab, what was the famous apple tree at the Battle of Hastings? At the center of the action in early medieval maps purporting to show the site of the battle is an apple tree with large, bright red fruits. In the grim words of *The Anglo-Saxon Chronicle* (Worcester Chronicle, British Library, Cotton manuscript),

> Then Count William came from Normandy to Pevensey on Michaelmas Eve [28 September 1066] and as soon as they were able to move they built a castle at Hastings. King Harold was informed of this and he assembled a large army and came against him at the hoary apple-tree, and William came against him by surprise before his army was drawn up in battle array. But the king fought hard against him, with the men who were willing to support him, and there were heavy casualties on both sides. Then King Harold was killed and Earl Leofwine his brother, and Earl Gyrth his brother, and many good men, and the French remained the measters of the field.

And the red apples of that lichen-encrusted ("hoary") tree would have fallen among the bodies of the nobility of Anglo-Saxon England. By William's order the bleached bones of the dead lay unburied on that hill well into the 12th century.

Despite the savage Norman repression of the Anglo-Saxons in places, there was a little light in the gloom: "amongst the followers of William there was a lady of the name of Mabilia. She fixed her residence in Kent, at one of the places where apples, it would seem, were already plentiful, and commending herself to the people by her virtues, became known as Mabilia d'Appletone, or Mabilia of the apple orchard. Her descendants, the Appletons of Kent and the adjoining counties . . . still, after 800 years, cling faithfully to their ancestral soil. The heraldic crest became an apple-bough, with leaves and fruit, and continues such to the present day" (Grindon 1885). The family name Appleton (cognate with *ton* are *tuin,* Dutch for garden, and *Zaun,* German for fence) is still common, along with street names and precinct names in the area of Canterbury, Kent, and has outlived the local family, whose name derives from *Mabilia,* the Maberleys and Mabberleys of that county, later Maverleys and Moverleys (Maberly Family 2005).

The sweet apple, *Malus pumila,* was brought across the English Channel and helped to colonize the British Isles at the time of the Roman invasion or, as argued, a little earlier. But what about the rest of Europe? The penetration was surprisingly patchy. In all those countries that came in part or wholly under Roman rule, the apple flourished. But under Celtic influence it does not seem to have extended as far as Scandinavia.

It seems that the delights of the pippin and reinette were very late in entering Sweden, for example. The Thirty Years' War (1618–1648) convulsed mainland Europe. Sweden, particularly after 1630 under Gustavus II Adolphus (d. 1632), Queen Christina, who reigned 1632–1644, and General Lennart Torstensson, raised large mercenary armies to fight for the Protestant cause. At the end of the war when Swedish armies returned from the Continent, they were said to have brought back large sweet apples in their knapsacks. It is supposed that the soldiers looted these exceptional fruits from orchards in southern Germany. The seeds, it can be assumed, germinated well in the harsh Scandinavian winters and formed the feral populations that are now widespread in southern Sweden. Indirect confirmation of a lack of sweet apples in Scandinavia before that time comes from the distribution of apple place-names in Britain. Such place-names are common in the regions overrun by the Saxons but rare in the northern and western areas settled by the Norwegians.

Reception of the apple in the West

In this book we argue that what we now know as the domestic apple in the West is an alien that came from Inner and Central Asia, uncontaminated by hybridization, as an adaptable and apparently universally welcome economic migrant that entered western Europe via the Persian empire, Macedonian and Hellenic Greece, and imperial Rome, and possibly but conjecturally via Celtic terrestrial and maritime migrations.

Yet for reasons that are not at all clear, unlike almost every other food source, whether native or alien, the apple entered, in every country into which it was brought, and almost immediately, the realms of the names of village and private residence, poetry, prose, the visual arts, language (including aphorisms), music (including carols), mythology, and philosophy. The apple even succeeded in being hijacked for the names given to later, imported, unrelated fruits and vegetables, for example, *pomme de terre* (French), *Erdapfel* (Austrian but not usually modern German), and *aardappel* (Dutch) for potato (*Solanum tuberosum*); *pomodoro*, Italian for tomato (*S. lycopersicum*); Punic apple or pomegranate, "the apple with many seeds" (*Punica granatum*); pineapple (*Ananas comosus*); and star apple (*Chrysophyllum cainito*). What were the reasons for this unparalleled emotional power and iconic popularity?

And there was soon to be more than just names of places and other fruits and vegetables—in fact, there was an explosion of literature and illustration inspired by the apple. The sweet apple, unlike almost every other alien fruit, appears to have received a degree of adulation in every form in the West, unparalleled except possibly for the rose (Langley 1729, Hooker 1818, Bunyard 1933, H. V. Taylor 1948, Morton Shand 1949, McLean 1981, Morgan 1982, Bultitude 1983, Ward 1988, Roach 1985, Petzold 1990, Wasserman et al. 1990, Morgan and Richards 1993, R. Palmer 1996, Browning 1999, Pollan 2001, Palter 2002, Clark 2003).

Eve and Aphrodite carry the apple in their hands; the serpent in Eden guards it; it is celebrated by Solomon; Ulysses yearns for it in the garden of Alcinous. In *The Odyssey*, Ulysses reminds his aged father that when as a little boy he had given him, for his own garden, "thirteen pear trees, and ten apple-trees, and forty fig-trees—I asked each of thee, being a child, following thee through the garden, and thou didst name and tell me each" (Grindon 1885). About 2,600 years ago Sappho wrote, "As the sweet-apple reddens on the bough-top, on the top of the topmost bough; the apple-gatherers forgot it—no, they did not quite forget, but could not reach so far" (Page 1955; the allusion is that the girl, like the apple, remains intact despite the zeal of her pursuers).

About A.D. 50, Petronius Arbiter, in Rome, wrote a poem to his mistress: "You send me golden apples, my sweet Martia, and you send me the fruits of the shaggy chestnut. Believe me, I would love them all; but should you choose rather to come in person, lovely girl, you would beautify your gift. Come, if you will, and lay sour apples to my tongue, the sharp flavor will be like honey as I bite. But if you feign you will not come, dearest, send kisses with the fruit; then gladly will I devour them." Keeping in mind the example of the apple in the Garden of Eden, however, it is possible that these golden apples might really be quinces (*Cydonia*) or even *Citrus*.

The Druids are reported to have selected their divining rods from the apple tree. The prophet Muhammad inhales eternal life through the scent of an apple that an angel had brought him. In the mythology of *The Arabian Nights* the story is told of one Prince Ahmed who purchased a magic apple in Samarqand that would cure all diseases (Room 1998). At least the geography is accurate with Samarqand one of the first major cities the migrating apple would have reached on its westerly journey. Many painted it. Magritte in particular loved it, and Walt Disney, in *Snow White and the Seven Dwarfs* (1937), imparts to the apple an evil symbolism unforgettable by any child of that generation, and many since.

Shakespeare (1564–1616) often mentioned apples. In *Henry IV, Part 1*, Act 3, Scene 3, Falstaff says to Bardolph, "Why, my skin hangs about me like an old lady's loose gown; I am withered like an old apple-John." *Henry IV, Part 2*, Act 5, Scene 3, in the garden, Shallow says, "Nay, you shall see mine orchard, where, in an arbour, we will eat a last year's pippin of my own graffing, with a dish of caraways, and so forth"; and later, Davy, setting them before Bardolph says, "There's a dish of leather-coats for you," In *Love's Labours Lost,* Act 4, Scene 2, Holofernes says, "The deer was, as you know, sanguis, in blood; ripe as a pome-water, who now hangeth like a jewel in the ear of caelo, the sky," and in Act 5, Scene 2, Berowne says to Boyet, "And laugh upon the apple of her eye?" In *The Merry Wives of Windsor,* Act 1, Scene 2, Sir Hugh Evans addresses Simple, "I will make an end of my dinner; there's pippins and cheese to come."

From the pen of Michael Drayton (1563–1631) came a catalog of apple varieties in his *Poly-olbion,* Song 18:

When as the pliant Muse, straight turning her about,
And comming to the Land as *Medway* goeth out,
Saluting the deare soyle, o famous *Kent,* quoth shee,
What Country hath this Ile that can compare with thee,
Thy Conyes, Venson, Fruit; thy sorts of Fowle and Fish
Which has within thy selfe as much as though canst wish?
As what with strength comports, thy Hay, thy Corne, thy Wood:
Nor any thing doth want, that any where is good.
Where *Thames*-ward to the shore, which shoots upon the rise,
Rich *Tenham* undertakes thy Closets to suffize
With Cherries, which wee say, the Sommer in doth bring,
Wherewith *Pomona* crownes the plump and lustfull Spring;
From whose deepe ruddy cheeke, sweet *Zephyre* kisses steales,
With their delicious touch his love-sicke hart that heales.
Whose golden Gardens seemeth *Hesperides* to mock:
Nor there the Damzon wants, nor daintie Abricock,
Nor Pippin, which we hold of kernell-fruits the king,
The Apple-Orendge; then the savory Russetting:
The Peare-maine, which to *France* long ere to us was knowne,
Which carefull Frut'rers now have denizend our owne.
The Renat: which though first it from the Pippin came,
Growne through his pureness nice, assumes that curious name,
Upon the pippin stock, the Pippin beeing set;
As on the Gentle, when the Gentle doth beget
(Both by the Sire and Dame beeing anciently descended)
The issue borne of them, his blood hath much amended.
The Sweeting, for whose sake the Plow-boyes oft make warre:
The Wilding, Costard, then the wel-known Pomwater,

And sundry other fruits, of good, yet severall taste,
That have their sundry names in sundry Countries plac't:
Unto whose deare increase the Gardiner spends his life,
With Percer, Wimble, Sawe, his Mallet, and his Knife;
Oft covereth, oft doth bare the dry and moystned root,
As faintly they mislike, or as they kindly sute;
And their selected plants doth workman-like bestowe,
That in true order they conveniently may growe.
And kills the slimie Snayle, the Worme, and labouring Ant,
Which many times annoy the graft and tender Plant:
Or else maintaines the plot much starved with the wet,
Wherein his daintiest fruits in kernels he doth set:
Or scrapeth off the mosse, the Trees that oft annoy.

Andrew Marvell (1621–1678) wrote in *The Garden,* verse 5:

What wond'rous Life in this I lead!
Ripe Apples drop about my head;
The luscious clusters of the Vine
Upon my Mouth do crush their Wine;
The Nectaren, and curious Peach,
Into my hands themselves do reach;
Stumbling on Melons, as I pass,
Insar'd with Flow'rs, I fall on Grass.

In more homely language the apple also had its place in the British Isles, most charmingly, and wisely, perhaps, in the exquisite Welsh proverb, "A seed hidden in the heart of an apple is an orchard invisible." It should not be a surprise that the apple, with its track record, was chosen as the focus for the medieval story of William Tell, wherein a villager seeks to achieve his freedom from harsh feudal edict by shooting an apple with his crossbow from the head of his son. Nor, too, that Isaac Newton was recorded as observing the fall of an apple in the garden of his mother's house. As any advertising executive would confirm, if you have an outstanding icon, exploit it to the full regardless of context, relevance, or historical accuracy. This vast tapestry of erotic imagery, sexual analogy, symbolism, and history is explored and analyzed in scholarly detail in the beautifully illustrated *La Pomme* (Wasserman et al. 1990). And astonishing in the eclecticism displayed, an impressive collection of poems, apocrypha, music, and public records of the apple in every guise has been gathered together in *Ripest Apples* (R. Palmer 1996). A similar compendium of the literary riches, ranging from Homer to *Sunset Boulevard*, can be found in *The Duchess of Malfi's Apricots and Other Literary Fruits* (Palter 2002).

The image of the apple

During the period of great church building in Europe, carver Gislebertus was at work on the cathedral of St. Lazarus in Autun in Burgundy. Over a lintel at the north door he carved in stone a beautiful image of Eve, decorously shielded by a vine branch, picking an apple. The apple, with its large fruits, short pedicels, and broad leaves with well-marked veins is obviously *Malus pumila* and not any form of northern wild crab. The belief that the sweet apple grew in the Garden of Eden, the land between the rivers Tigris and Euphrates, was completely erroneous. But most significantly, this beautiful carving indicates that the essential features of the sweet apple were clearly recognized at this early date.

The apple in devotional pictures, frequently of mythical depictions of the Garden of Eden, begins to appear in western European art from the early Middle Ages. Indeed, "the apple of one's eye," referring to the pupil, anciently thought to be round and solid like an apple, but now meaning a much loved person, comes

Eve from the cathedral at Autun, France, by Gislebertus, the carving dated to the year 1135.
Reproduced by kind permission of the curators of the Musée Rolin d'Autun.

from Deuteronomy 32 (Room 1998). The myth that apples grew in Mesopotamia has proved difficult to eradicate. The simple cold-chill requirements for apple seed germination ought to refute that belief but have failed lamentably. The etymology of the word for apple provides no help. The core word *mal* seems simply to mean a round, probably fragrant fruit, and it has been argued (Church 1981) that the Eden apple was *Strychnos nux-vomica.*

In most early depictions, the apple, though obviously *Malus pumila,* is either stylized beyond cultivar recognition or rendered almost unrecognizable by the crude technology of early woodcuts. But within a few years, astonishingly and seemingly uniquely, the humble apple became a portrait subject in its own right. To be fair, the Romans, in their exquisite murals at Pompeii, came close to verisimilitude (Wasserman et al. 1990). By the first decades of the 16th century and close in time, in Lucas Cranach's *Cupid Complaining to Venus* and Jan Gossaert's *The Virgin and Child,* the apple is very carefully and individually painted. The Gossaert apple closely resembles the well-known early French variety 'Api Étoilée', a strongly five-ribbed, dessert apple first described by Jean Bauhin (1598), a physician and botanist attached to the court of the duke of Württemberg (Bugnon 1995). The apple painted in Jan Van Huysum's *Dish of Fruit* (about 1720) is seemingly identical to Mary Martin's 20th century painting of 'Queen' (Spiers 1996). Just how esteemed the apple had become can be understood by a glance at the walls of any major art gallery in the world. What other fruit or vegetable—orange? banana? grape? potato?—has been accorded such artistic respect, by such a range of the world's outstanding artists? Magritte in particular seems to have been obsessed by the apple:

Lucas Cranach (1472–1553), *Cupid Complaining to Venus,* National Gallery, London

Jan Gossaert (about 1478–1532), *The Virgin and Child,* National Gallery, London

Dosso Dossi (Giovanni Luteri; about 1490–1542), *Allegory of Hercules—Sorcery,* Galleria degli Uffizi, Florence

Giuseppe Arcimboldo (about 1530–1593), *Four Seasons, Autumn,* Pinacoteca Civica Tosio-Martinengo, Brescia

Joachim Beuckelaer (about 1533 to about 1574), *Earth,* a Netherlands fruit market scene, from a series of four paintings: *Air, Earth, Fire,* and *Water,* National Gallery, London

Juan Sánchez Cotán (1560–1627), *Still Life with Game Fowl, Fruit and Vegetables,* Museo del Prado, Madrid

Georg Flegel (1566–1638), *Man and Woman Before a Table,* private collection

Caravaggio (Michelangelo Merisi; 1571–1610), *Basket of Fruit,* Pinacoteca Ambrosiana, Milan

Nathaniel Bacon (1585–1627), *The Cookmaid with Still Life of Vegetables and Fruit,* Tate Gallery, London

Balthasar van der Ast (1593–1657), *Basket of Fruits,* National Gallery of Art, Washington, D.C.

Louise Moillon, (1610–1696), *The Merchant of Fruits and Vegetables,* Musée du Louvre, Paris

Gerard ter Borch (1617–1681), *The Apple Peeler,* Kunsthistorisches Museum, Vienna

Jan Van Huysum (1682–1749), *Dish of Fruit,* Teylers Museum, Haarlem

Jean-Baptiste-Siméon Chardin (1699–1779), *White Teapot with White and Red Grapes, Apples, Chestnuts, Knife and Bottle,* private collection

N. F. Gilet (about 1757), *Paris—The Shepherd, with an Apple,* Musée du Louvre, Paris

Luis Mélendez (1716–1780), *Still Life with Fruit, Cheese and Containers,* Museo del Prado, Madrid

Samuel Palmer (1805–1881), *The Magic Apple Tree,* Fitzwilliam Museum, Cambridge, England

Jean-François Millet (1814–1875), *A Peasant Regrafting His Fruit Trees,* Bayerische Staatsgemäldesammlungen, Munich

William Holman Hunt (1827–1910), *The Hireling Shepherd,* Art Gallery of the city of Manchester

Édouard Manet (1832–1883), *Still Life—Fruit on a Table,* Musée d'Orsay, Paris

Edward Burne-Jones (1833–1898) and J. H. Dearie, *Pomona,* silk and wool tapestry, Victoria and Albert Museum, London

Henri Fantin-Latour (1836–1904), *Flowers and Fruit,* Musée d'Orsay, Paris

Paul Cézanne (1839–1906), *Still Life with Apples,* Fitzwilliam Museum, Cambridge, England; *Apples and Cake,* private collection (sold 2005 in New York)

Henri Matisse (1869–1954), *Apples,* Art Institute, Chicago

René Magritte (1898–1967), *Ceci n'est pas une Pomme,* private collection

Francine Van Hove (b. 1942), *Les Pommes,* Galérie Alain Blondel, Paris

By the time of that outstanding Huguenot woman painter of the 17th century, Louise Moillon, it was possible to identify, in a market stall, a great range of apples, including fully flushed desserts and russets, indistinguishable in size, quality, and range from a modern fruit display. From the very earliest depictions it would seem that the apple had become an erotic icon. Large, round fruits were a consistent symbol allied to the nakedness of the images in the Garden of Eden. But even when scenes of the garden began to disappear from the artists' studios, the apple retained its association with sexuality in general and the female breast or buttock in particular (Palter 2002). This imagery was developed overtly in Roman mosaics, in Dossi's *Allegory of Hercules—Sorcery*, Van Hove's *Les Pommes*, and indirectly in Holman Hunt's *The Hireling Shepherd*.

And what pomological library could be without *The Herefordshire Pomona* (Hogg and Bull 1876–1885), *The Apples of England* (H. V. Taylor 1948), *Apples* (Bultitude 1983), *The English Apple* (Sanders 1988), *The Book of Apples* (Morgan and Richards 1993), or Mary Martin's lifetime dedication to the cider apple in *Burcombes, Queenies and Colloggetts* (Spiers 1996)?

The imagery also survives in an unexpected way in that the familiar name for New York City is "The Big Apple" (Society for New York City History 1995). Regularly attributed to early jazz celebrities or sports heroes, the name actually derives from a furnished bordello in a building still standing at 142 Bond Street, a business established in the early 19th century by an emigré from France, Mlle Evelyn Claudine de Saint-Évremond. Soon known locally as Eve, she in consequence referred to her ladies as "my irresistible apples," while her patrons spoke of visits as "having a taste of Eve's apples." With time, apple catchphrases such as "The Apple Tree'" and "The Real Apple" were used as names for the city itself, and by 1907 "The Apple" or "The Big Apple" had passed into general use—without the original connotations. Indeed, it was to be completely sanitized by the Apple Marketing Board in upstate New York. Alarmed at declining sales of the fruit, the board initiated a campaign to reverse the trend and promoted such wholesome phrases as "An apple a day keeps the doctor away" and "as American as apple pie." How Mlle Evelyn and her patrons would have laughed!

The apple in Germany

In the development of grafting and from the achievements of medieval artists, the juggernaut of apple adulation gathered speed. The first serious representational, unstylized artwork depicting the apple dates from the early 1500s, and the poets and writers of prose were not far behind. From Hieronymus Bock (1498–1554), a German writing as early as 1546 in Latin, comes the following:

How richly endowed is Nature, mother of all things, and how full of marvels, is abundantly shown by all her products, but by nothing so much as by the diversity and almost unlimited number of varieties of apple. Who could claim to describe the shape, smell, and flavor of all of the apples that occur in Germany alone, to say nothing of those of other regions? No one, not even Cloatius himself [a writer of the 1st century B.C.] who, according to Ruellius's *On the Nature of Trees,* [Ruel 1536] Bk. 1, Chapter 97, listed some 20 varieties of apple, could so much as assign names to all the individual kinds of apple. Though as I have said Nature has been most lavish in the creation of fruit and especially of apples, I will endeavor to give a brief resumé of at least the commonest apples of Germany. There are cultivated and there are wild apples. Under these two headings are grouped an almost unlimited number of varieties of apple. Some kinds are large, some small. Some kinds are round, others oblong. Some kinds are sour, others sweet. Some kinds are early ripening, others late ripening. Some kinds are white-skinned, others yellow-skinned. Some kinds have a little redness, others a lot. Some kinds are red only on the outside, others are red inside as well, others again are splashed with oblong red marks shaped like weals.

But varied and diverse as these varieties of apple are, they all belong to much the same trunks, the same wood, the same branches, the same leaves, and closely similar flowers. They are all as quick to grow as they are quick to perish. Their trunks are covered with a bark that is scabrid and grayish externally and internally is of a waxen color; it is used to make a yellow dye.

The apple likes a fertile, rich, cool, damp soil. Its leaves are rounder than those of the pear and larger than those of the quince. It flowers later than the pear, at the beginning of May. The blossom is pure white in some varieties, pink-tinged in others. Apples ripen at different times. Some ripen early around St. John's day. Others ripen in August. The latest apples ripen in autumn. Apples reveal their ripeness by the blackness of their seeds.

Old writers directed that the apple should be sown at two seasons of the year, in spring and in autumn. On this subject, see Columella in his book on trees.

Care must be taken to protect the roots of apples from worms. This can be done if we follow the example of the ancients and pour pig's dung mixed with human urine over the roots. Human urine will be of exceptional benefit to any trees with worm-eaten roots over which it is poured.

PLATE 14. The jagged mountains of the Tian Shan from the north, on the Kazakh-Kyrgyz border. Reproduced by kind permission of Dr. Hartmut Bielefeldt.

PLATE 15. The Tian Shan near Tian Chi, Xinjiang Uygur, China. Note the fir trees, *Picea schrenkiana*. Reproduced by kind permission of Dr. Alan Whittemore.

PLATE 16. Fresh exposed limestone cliff faces with caves in the mountains of the Tian Shan north of Jalal-Abad, Kyrgyzstan, September 2001. New forest is forming at the base of these uplifting cliff faces.

PLATE 17. The fruit forest in the Dzungarian Alatau, Kazakhstan, September 1999.

PLATE 18. Salt deposits on the steppe northeast of Almaty, Kazakhstan, September 1999.

PLATE 19. Honeybee colonies (*Apis mellifera*) in the Ili Valley of China in September 2000 but brought down from the mountains a year before.

PLATE 20. Very large *Malus pumila* tree, probably more than 12 m (39 feet) tall, in the Dzungarian Alatau, Kazakhstan, September 1999.

PLATE 21. A range of fruits gathered in a single day in September 1999 in the fruit forest of the Dzungarian Alatau, Kazakhstan.

PLATE 22. A yurt in Kazakhstan: the all-season, demountable, wool-felt tent, September 1999.

PLATE 23. Inside a yurt, showing the light, willow-frame construction; Dzungarian Alatau, Kazakhstan, September 1999.

PLATE 24. Trophy bearskin (*Ursus arctos*) in a yurt near Almaty, Kazakhstan, September 2002.

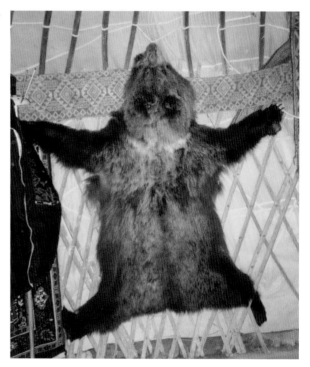

PLATE 25. Bear droppings with a rich burden of apple pips in the Dzungarian Alatau, the mountain range between eastern Kazakhstan and northwestern Xinjiang Uygur, China, September 1999.

PLATE 26. Apple wood collected for winter fuel near Jalal-Abad, Kyrgyzstan, September 2001.

PLATE 27. Roadside retailing of seedling apples near Almaty, Kazakhstan, September 1999.

PLATE 28. Horse and camel, Kazakhstan, September 1999.

PLATE 29. Kyrgyz citizens near Jalal-Abad, Kyrgyzstan, September 2002.

PLATE 30. Drying fruit on a stretch of road near Turpan, Xinjiang Uygur, China, September 1999.

PLATE 31. A modern paradise garden with corrugated iron and barbed wire near Jalal-Abad, Kyrgyzstan, September 2001.

PLATE 32. *The Caravan at Rest by the Coffee House near Scalanova in Asia*, camels and horses in a caravan separately corralled near Kusadasi, a seaport opposite the island of Sámos in western Turkey, 1794. Drawing by Ferdinand Bauer (see Lack with Mabberley 1999), reproduced by kind permission of the Director of Library Services, University of Oxford.

PLATE 33. Cleft grafting in a Roman mosaic, *Mosaïque du Calendrier Rustique.* Photograph by Erich Lessing, Art Resource, New York. Reproduced by kind permission of the curators of the Musée des Antiquités Nationales in Saint-Germain-en-Laye, France.

PLATE 34. Fruit pressing in a Roman mosaic, *Mosaïque du Calendrier Rustique.* Photograph by Bridgeman-Giraudon, Art Resource, New York. Reproduced by kind permission of the curators of the Musée des Antiquités Nationales in Saint-Germain-en-Laye, France.

PLATE 35. Apples hanging on a tree after overwintering there; Oxford, England, 14 March 2002.

PLATE 36. Natural grafting of beech (*Fagus sylvatica*) in a formerly pleached hedge, with Sarah Juniper, near Uley, Gloucestershire, England, March 2002.

On the Names

The Germans have a large number of different kinds of apples and have bestowed on them a correspondingly large number of different names. I will give a brief summary of these.

The first apples to ripen with us are called Sanct Johans Opffel [St. John's Apple or apple-Johns (Room 1998), allegedly ripening by 24 June, St. John the Baptist's day, the *mele di San Giovanni* of present-day Istria, Croatia], in Latin *Praecocia Male* [early apples].

Apples which come in August, and which are covered with red marks like weals and are sweet, we call Augstoppfel. In Greek they are called *Glycymele* [sweet apples].

Apples which ripen in autumn and which are either white-skinned, or yellow-skinned, or have flushes of color, but in all cases are sweet, we call Würzoppfel [spice apples], that is, aromatic apples, by reason of their delicious smell and flavor. Others call them Capendua. Apples which are large, white and sweet, and which have their seeds rattling in their heart at maturity, are called, by Ruellius, Passipoma.

Other apples, equally large and similarly sweet but red-spotted, are called by the Germans Schragenopffel or Hergotsoppfel.

Other sweet, white-skinned apples occur which are turbinate and oblong, called 'Orthomastic' [high-breasted] apples, in German Stromelting [striped apples?] and Genszopffel. There are also apples called Paradisian, in German Paradeiszopffel. There is also an unlimited number of sour apples. The commonest of these are designated in Germany by the following names: Kolopffel: Weinopffel [wine apples], Heissling / Hermelting / Stremling [striped apples], Speierling / Heimelting / Frawenopffel [ladies' apples]. Last come crab apples, in German Holzopffel [wood or woody apples], which are themselves of more than one kind. Of these varieties, the three I like best are the apples they call the Graw Würzopffel [gray spice apple], next Gälen Hermelting, finally those I have referred to as Hermelting.

Internal Application of Apples

An excellent electuary is made from vinous apples in the same way as quinces, which is both pleasing to the palate and very suitable in burning fevers, for it quenches the thirst and stimulates the flagging appetite, in a manner more agreeable than any other preparations of pharmacies.

> Sour apples restrain the action of the bowels, check the urine, stop vomiting and belching, a property belonging most particularly to crab apples. Sweet apples are said to release the bowels and expel worms.

There are inconsistencies in the spelling of certain German words, for example, *Apffel*, *Opffel*, and *Oppfel*, (modern German is *Apfel*), but this is not surprising in the early 16th century. In German-speaking countries between 1750 and 1800 there were more than 120 distinct books on the subject of the apple (Janson 1996). Many of these contained plagiarized medieval nonsense along the lines that fresh ox blood poured over the roots of an apple tree would turn the fruits red. But the sheer number is an illustration of the fascination with the fruit.

That puzzling word *reinette*

Among the outstanding grafted dessert varieties of northern Europe and Britain are many with names including the word *reinette*, for example, 'Ananas Reinette', 'Orléans Reinette', 'Reinette du Canada', 'Reinette Rouge Étoilée', 'Heusgen's Golden Reinette', 'Luxemburger Renette' (sic), and 'Reinette d'Obry', the cultivars predominantly but not exclusively French. The first use of the name seems to be about 1540 with the French 'Reinette Franche'. Then came a cluster of extant cultivars in the early 17th century with the French 'Reinette de Mâcon', the English 'Golden Reinette' about 1650, and the still widely cultivated French 'Reinette Grise Ancienne' about 1653. It seems most likely that *renatus*, from the Latin for a rebirth, became corrupted into *reinette* and caused endless confusion by its possible misconstruing into having some relationship with the French *reine*, queen, as "the little Queen's apple." Langford (1681a) wrote, "thence had its name in a manner of being born again."

One interpretation is that at least some reinettes are sports, displaying distinctive and attractive differences, from a well-established cultivar. 'Cox's Orange Pippin' is particularly inconstant in this respect, generating a clutch of minor fruit mutants. Having the supporting and well-tested genotype of the parent cultivar, they have been selected through the centuries as successful cultivars in their own right. Confirmation of this appears in a line in Drayton's *Poly-olbion:* "The Renat: which though first it from the Pippin came."

The textbook and the apple

Grafting apart, the general cultivation of the orchard was already well catered for by the end of the 17th century. Worlidge (1669), Meager (1670), Langford (1681a, b), and "A Lover of Planting" (1685) gave general guidance. Beale (1653), Austen (1657), Rea (1665), and Venette (1685) gave comprehensive and accurate directions for pruning. The same was true, in the vernacular, across the whole of Europe, particularly from the publication of Johann Domitzer's (1531) *Ein Neues Pflantzbüchlin* and the translation of Petrus Lauremberg's (1631) *Horticultura* from Latin into German (see Janson 1996).

Hieronymus Bock extolled the virtues of the apple and demonstrated the range of cultivars but, apart from some doubtful manuring information, gave us little information on how to grow the alien fruit. But roughly from the second quarter of the 16th century, the earliest period when textbooks began to emerge, publications devoted to the apple and, to a lesser extent, the pear tumbled off the presses (Janson 1996). Examined in detail—only fragments remain of some—they are found to contain a great deal of recycled myth and misleading information. But around the same time as Bock, Johan Domitzer (1531), John Fitzherbert (1548), Thomas Hill (1563), Leonard Mascall (1572), and John Gerard (1597) all provide high-quality information, in the vernacular, on the techniques of grafting (Chapter 4).

By the 17th century there is not only a literary outpouring of adulation as well as good information, but also lists of the most suitable cultivars: William Lawson (1618), Gervase Markham (1625, 1640), Ralph Austen (1657, 1658, 1676), John Evelyn (1664), John Rea (1665), Francis Drope (1672), John Worlidge (1676), Thomas Langford (1681a, b), "A Lover of Planting" (1685), an anonymous writer who copied some from Langford (Juniper and Juniper 2003), George London and Henry Wise (1699a, b), and Samuel Collins (1717). On the Continent there was, for example, Jean de La Quintinye's (1690) *Instruction pour les Jardins Fruitiers . . .* , and numerous apple cultivar engravings from Frankfurt in *Dendrographias* by Joannes Jonston (1662), a well-traveled Pole of Scottish extraction. And in the New Plantations, that is, North America, there were lists of fruits to be planted in orchards by, for example, William Hughes (1672).

Many of these texts were translated backward and forward between languages, and the illustrations therein traveled with them. The influential and well-illustrated *Le Jardinier du Pays-Bas* by Jan van der Groen (1669, 1681) was published in Dutch, French, and German versions. Most of its illustrations were copied from Lauremberg's (1631) *Horticultura* and other earlier works. Charles Cotton's (1675) *Planters Manual* is not actually an original work, as was claimed, but a translation of *Instructions pour les Arbres Fruitiers* by, it is generally written, Robert Triquet (Janson 1996 has this as the old French form "Triquel"), but likely not even by him—more plagiary—but by François Vautier, physician to Louis XIV (Bunyard 1918). John Evelyn, finding time to spare apart from writing about salad crops (Evelyn 1699), translated the works of at least three French pomologists: Jean Baptise Le Gendre, Nicolas de Bonnefons, and Jean de La Quintinye. Some works went from French to English, Italian, Dutch, and German, in any order. Often there was little or no concern for originality, and it becomes a major labor for bibliophiles to decide whose text, whose illustrations, and in what language the originals appeared. Subsequent to the enactment of copyright law in 1709, the first completely original text in this genre is probably Philip Miller's *Gardeners Dictionary* of 1731.

There was no letup, although rather more originality, in the 18th century, particularly when it began to be possible to produce illustrations in color, albeit hand-colored copper engravings, such as Johannes Knoop's (1758) sumptuous *Pomologia*. Stephen Switzer (1724) gave comprehensive instructions, particularly on grafting, rootstocks, and pruning. Batty Langley (1729; and see Henrey 1975) combined exotic garden architecture with a comprehensive knowledge of fruit cultivars; some versions of his texts, too, were also expensively hand-colored. Thomas Hitt's (1755) *Treatise of Fruit-Trees* went through no fewer than five editions. Thomas Barnes (1759) was influential, particularly on root cuttings, and the prolific Thomas Andrew Knight (1797, 1811) continued to spread the pomological gospel.

From the early years of the 19th century, the floodgates were opened. Pomonas and other texts, in all the major vernacular languages, poured forth. Probably the first and the most distinguished full-color-printed fruit book was a posthumously published edition (1807–1835) of the work of that indefatigable fruit collector, Henri Louis Duhamel du Monceau (1700–1782), which was followed by Hooker (1818; and 1989), Loudon (1844), Maund (1845–1851), Hogg (1851, 1884), Hogg and Bull (1876–1885), and the renowned John Scott's (1873) *The Orchardist* (also see Raphael 1990 and Janson 1996 for outstanding examples of beautifully produced, widely printed adorations of the apple). Hand coloring was cripplingly expensive, and Robert Hogg was forced to concede that such works were "from their great cost to be regarded more as works of art, than of general utility."

From the end of the 19th century, as full-color printing became widely available, albeit erratically, the pomona became more common. Perhaps the most effusive and elegant pomologist was Edward Bunyard (1878–1939), who encapsulated the whole mystique of apple appreciation in his fulsome *A Handbook of Hardy Fruits More Commonly Grown in Great Britain* (1920) and *The Anatomy of Dessert* (1933), wherein he wrote about the acidity, aroma, flavor, and texture of many of the finest apples.

The apple in folklore and other stories

The extent to which the apple and its many admirers have entered the folklore and history of nations can be illustrated by five thumbnail sketches. It is difficult to think of any other fruit or food that has attained such prominence (Bazeley 1991), and the omissions here will probably cause friction. There are many other stories of apples with enduring reputations (Morgan and Richards 1993, Spiers 1996, Copas 2001).

Newton's apple

Isaac Newton was born at Woolsthorpe Manor in Lincolnshire, England, in 1642. There can be no doubt that Newton was a careful observer of many apects of natural history, and the fall of an apple from a tree might well have been noted, but in the story of Newton's apple the penetration of the iconic fruit into folklore can be seen (Tallents 1956). The story first appeared in 1727, the year of Newton's death, in a popular account of his work by Voltaire. Recording a conversation with Catherine Barton, Newton's niece, confidante, and housekeeper, Voltaire noted, "One day in 1666, Newton having returned to the country [he had come back to avoid the plague] and seeing the fruits of a tree fall, fell . . . into a deep meditation about the cause that thus attracts all bodies." It is now thought probable that Newton spoke of the fall of the apple not in 1666 but toward the end of his life, to add color and a focus to his discovery of the laws of gravity. He succeeded beyond his wildest imagination.

Newton's apple has been identified as 'Flower of Kent', first recorded by Parkinson (1629) and still grown in specialist collections. It has been argued that Newton's own tree still exists. It was first described in 1806 and drawn in 1820, and these pieces of evidence point to a gnarled and twisted specimen still alive on the grounds of the manor. Carbon-14 analysis of a sample from the upper, younger branches gave a date in the 1700s. However, the original tree, supported by props, is stated to have died in 1814, with part of its wood used to make a chair in the Woolsthorpe Library (Morgan and Richards 1993). The possibility of self-layering makes these two statements not necessarily contradictory. The apple, in this context, even managed to get onto a British 18-pence postage stamp commemorating the 300th anniversary of Newton's *Principia*.

'Norfolk Beefin'

Just a very few individual apple cultivars have entered folklore for their culinary excellence (Ward 1988). 'Norfolk Beefin' (Plate 10), the name suggested by John C. Smith to be derived from *peau fine*, French for "beautiful skin," which is certainly true of the apple, also known as the "beef apple," which may refer to its blood-red color, was a common feature of markets, particularly of East Anglia, in the 18th and 19th centuries. Parson James Woodforde, the diarist of East Anglia (1740–1803), wrote in April 1780, "sent Justice Buxton this morning a Baskett of my fine Beefans, a very fine kind of Apples," and a little later, "one apple from my Beefan tree weighed 13 oz. By which I got a Wager of Nancy of 0.6d." Beefins are typical phase 3 apples, hard enough practically to be fired from a cannon, and falling mostly unharmed from the tree onto the orchard floor.

Biffin is the Norfolk word for a cooking apple. At the end of the working day, when all the bread and confectionaries had been withdrawn from the now cooling, communal brick oven, a tray of biffin apples was prepared. "Take an oven tray," read the instructions of the time (Beresford 1924, Morgan and Richards 1993), "cover it with wheat straw on which you place the apples, then cover with another oven tray on which you set oven weights from seven to 11 pounds [3–5 kg] depending on the size of the tray. Then leave in the oven 40–48 hours." Their fame, in the cooked state, has persisted to this day. 'Blenheim Orange', 'Striped Beefin', or 'Herefordshire Beefin' will respond to this form of cookery, but reputedly no other.

'Bramley's Seedling'

'Bramley's Seedling' was raised in a cottage garden in Church Street, Southwell, Nottinghamshire, England, between 1809 and 1813 by Miss Betsy Brailford. The parentage of the seed is unknown, but the original tree is still alive. A handsome, brilliant red color mutant known as 'Crimson Bramley' arose in about 1913 in the Southwell area; this is also widely planted. Bramleys are triploid and therefore of little value as pollen parents as they do not produce viable pollen.

Although the original tree is well beyond the point at which good scion wood could be taken, cells have been isolated and propagated by tissue culture. These direct clonal derivatives have been distributed (Power and Cocking 1991). Many such cultures of existing clones, deriving as they do from apical cells, show a tendency to deviate from the character of the original cultivar, but this defect has not become apparent in the "original Bramley" clones.

In the nearly two centuries of its existence, 'Bramley's Seedling' has become one of the most highly appreciated of all culinary apples. The degree of adulation has reached the level of individual monographs (Merryweather 1992). The acreage devoted to its cultivation is certainly greater than that of any other cooking apple in Britain, and perhaps worldwide. Unlike many cultivars, such as the delicious 'D'Arcy Spice', which is more or less restricted to East Anglia, 'Bramley's Seedling' will seemingly grow almost anywhere.

The characteristics of a good cooker, an apple that needs cooking to be palatable, are a high acid and low tannin content. It is odd that a cooking apple, of which 'Bramley's Seedling' is the doyenne, scarcely exists in the repertoire of cultivars of other apple-growing countries, even where those countries can boast a list of named apples as great or greater than that in the United Kingdom. 'Rhode Island Greening', which probably arose as a chance seedling of unknown parentage at Green's End, Newport, Rhode Island, where an inn was kept by a Mr. Green, was well known and widely planted by the early 1700s. It was the basis of the original "momma's apple pie" (now a metaphor for sickly wholesomeness and complacency) and was grown by Thomas Jefferson at Monticello. Today it does not enjoy quite so elevated a status and is rarely available as a whole fruit in commercial quantities. It appears as a canned fruit and can still be bought fresh in the United States from a few specialist growers; a plant derived from it, just a sport (deriving from a somatic mutation in a growing tip) or a lucky seedling (nobody seems to know) in the western United States is known as 'Northwest Greening'. 'Bramley's Seedling', on the other hand, has been planted and has succeeded all over the world, a distinction achieved by no other cooking apple.

'Ribston Pippin'

Among the earliest apples to enjoy a national reputation, apart from 'Redstreak' for cider, was 'Ribston Pippin' (Plate 8). The word *pippin* (*pépin*, French for seed) simply denoted that the apple concerned was thought or known to have originated as a chance seedling. The first 'Ribston Pippin' grew at Ribston Hall near Knaresborough, Yorkshire, England, then owned by Sir Henry Goodricke (Simmonds 1946). The gardener, Robert Clemesha, or Clemensha, sowed a handful of seeds probably brought back as early as 1688 when Sir Henry and his bride spent part of their honeymoon near Rouen, France.

The original tree was blown over and damaged by cattle when about a century old, but it self-layered and, protected by a surrounding iron fence, survived until November 1928. By 1800, 'Ribston Pippin' was known not only all over Britain but was also exported abroad. The "Ribstone" was esteemed by Mrs. Beeton

(1861) as an excellent dessert. As is often the case with distinguished veterans of many species, its superb qualities were late in being recognized; it was not awarded a Royal Horticultural Society Award of Merit until 1962.

It came to have literary power as well. In *The Pickwick Papers*, Charles Dickens wrote, "a little hard-headed, Ribston Pippin-faced man." Hilaire Belloc observed in *The False Heart*, "I said to Heart, 'How goes it?'; Heart replied, 'Right as a Ribstone Pippin!' But it lied."

Darwin (1868) noted that 'Ribston Pippin' trees that had been exported to suitable, cool places in colonial India seemed to retain their original fruit character but assumed a fastigiate shape. One can only wonder at the difficulties of transporting whole trees, with their water-demanding root systems, halfway around the world on sailing ships.

'Blenheim Orange'

One of the best-known dessert cultivars, still widely grown, is 'Blenheim Orange'. Here the apple is coupled with imperial grandeur—still a winning marketing strategy. It is recorded that about 1740 an apple seedling, presumably from a discarded fruit, was found against the wall of Blenheim Park, the seat of the duke of Marlborough near Woodstock, Oxfordshire, England, by a Mr. Kempster, living in Old Woodstock. The Kempsters were a long-established family that included stonemasons and quarry owners, some of whom who had worked on building the palace.

The fruit attracted considerable interest on account of its size, beauty, color, and excellent eating quality. It became known as 'Woodstock Pippin' or 'Kempster's Pippin'. But in 1804, after a dish of these apples had been presented to the duke's head gardener, it was agreed that it might be called 'Blenheim Orange Pippin', and that, generally shortened to 'Blenheim Orange', is the name by which it is usually known. Indeed, apple cultivars have frequently been renamed to enhance their marketing potential. One of the best known is 'Winter King', raised in 1935 but renamed 'Winston' (Plate 9), after Churchill, in 1944. Unlike 'Ribston Pippin', 'Blenheim Orange' received early official recognition and was awarded the Banksian Silver Medal in 1820. It began to be grown all over the apple world but for obvious reasons was renamed in France, where it is known as 'Bénédictin'.

There is no illustration of Kempster's original tree, which died in 1853, or of the apples growing on it. One has to depend, therefore, upon apples with a reliable pedigree, derived by grafts from the original tree. By 1820, John Jefferies of Cirencester, Gloucestershire, was the first nurseryman offering 'Blenheim Orange' trees for sale. By 1822 or so they were available in a number of nurseries in London, and in the garden of the Horticultural Society at Chiswick. William Hooker produced a superb drawing of a 'Blenheim Orange' growing in the society's garden in 1821–1822 (Hooker 1989).

But are there any apples that can reliably be claimed to descend directly from Kempster's tree? One must approach this with circumspection since there is not the slightest doubt that somatic mutation (sporting) has taken place, as in 'Crimson Bramley'. Also, and as has been noted but is not understood, 'Blenheim Orange', though triploid, will set some seed, seedlings often coming more or less true (Robb-Smith 1956). In *Pomona Herefordensis* (Knight 1811) there is an illustration and description of a 'Blenheim Orange' very different from the apple illustrated by William Hooker some years later. That depicted by Hooker (1989) is now colloquially called 'Broad-Eyed Blenheim Orange' to distinguish it from classic 'Blenheim Orange'. It is the broad-eyed Blenheim that is ordinarily available from nurseries. But examples of what may be called the true Blenheim are still available from specialist nurseries.

'Blenheim Orange' trees are suspected to have given rise to a number of outstanding later cultivars. Among these are 'Bramley's Seedling' (about 1809), 'Cox's Orange Pippin' (about 1830), 'Annie Elizabeth' (about 1860), and 'Newton Wonder' (about 1870). Known 'Blenheim Orange' crosses (Robb-Smith 1956) are 'George Carpenter' (1902), 'Edward VII' (1903), and 'Howgate Wonder' (about 1915). Even if some of these paternity insinuations—and there are several more— are found by molecular methods not to be justified, 'Blenheim Orange' has proved to be a remarkable parent in spite of its irregular genetic constitution.

And some non-apples

Although "apple" has pervaded English and other languages with phrases and metaphors, for example, "upset the applecart" from as early as 1788, there has to be caution in that not all such "apples" are truly apples. The Classical apples of the Hesperides, often thought of as citrus fruits, are probably quinces (Mabberley 2004).

Again, although real apples have been used as flavorings and to scent products such as shampoo, they are often replaced by synthetic methyl acetate. Non-apples in English metaphor include "the apple of one's eye" (p. 140), which, like the apple of the Garden of Eden (and therefore "Adam's apple"), comes from the Old Testament, and apple word confusions include "apple-pie bed," with sheets so folded that a person cannot get his legs down between the sheets, which is believed to be derived from the French *nappe pliée*, folded cloth, while "apple-pie order" is perhaps from the French *cap-à-pied*, head to foot (Room 1998). And "apples and pears" is Cockney rhyming slang for "stairs."

CHAPTER 6

Apple migration across the seas

ALTHOUGH IT TOOK SOME 6,000 YEARS for the apple to be brought from the Tian Shan to western Europe, it took only 300 for it to reach all the other temperate parts of the world—at the hand of colonizing European powers. The durability and scurvy-deterring qualities pre-adapted the fruits to being transported. In its new habitats, *Malus pumila* exhibited a second unexplained release of variation, leading to the selection of some of today's most important commercial cultivars, notably 'Red Delicious' and 'Golden Delicious' in North America and 'Granny Smith' in Australia. Most of the old European cultivars were hardly satisfactory in the new environments, and selection has now moved to favor both low-chill and extended cold-chill cultivars in more extreme climates.

IT WOULD APPEAR that the apple took some 6,000 years to travel from the Tian Shan to the western fringes of Europe. Its subsequent colonization of both the western and southern hemispheres was far more rapid. From the end of the 16th century, European crops were pouring into the new colonies. Unlike almost every other fruit or other source of vitamin C, apples, or at least some of them, were easily transportable and therefore a boon to travelers. Cyder or cider (Chapter 7) was an excellent way of concentrating (and adding to) these virtues. Ralph Austen (1657) wrote in *A Treatise of Fruit-Trees*, "Cider is also of speciall use in *long voyages at Sea*."

Scurvy

The explorer and cartographer James Cook (1728–1779) is commonly supposed to have been the first significant advocate of what was generally called an anti-scorbutic, later identified as vitamin C, to ameliorate the disastrous effects of scurvy onboard ship. Cook is reported never to have a lost a man from scurvy. But in fact, empirical knowledge of the benefits of apples, along with other fruits and their products, were well known long before. For example, John Tradescant the Elder, gardener to Charles I, at the behest of Sir Dudley Digges on a sea voyage to northern Russia in 1618 (his personal mission to collect plants new to the British Isles), noted in his diary (Ashmole 1618, Leith-Ross 1984) that his ship rounded the North Cape on 6 July, and he landed at Archangel, in what is now Russian Europe, on 16 July. There he found a "bery growing lowe [possibly a cranberry, *Vaccinium uliginosum*, or it might have been the cloudberry, *Rubus chamae-morus*] which was eaten by the people . . . for a medssin against the skurbi," so immediately he collected "sume of the beryes to get seed whearof . . . and sent par to Robiens of Parris [Jean Robin, botanist to Henry III, king of France and founder of the Jardin du Roi, now the Jardin des Plantes]."

Cook is reported to have carried 'Hunt Hall', not listed now in the British national apple register (Janes 1998) nor by Morgan and Richards (1993) but still available from specialist growers in the Yorkshire area. Whaling ships of the 19th century, whose crews would be away for very long periods, stocked up at Whitby in northern England with robust (phase 3?) apples from Yorkshire orchards as well as barrels of cider (Twiss 1999). Among the apples would probably have been the well-known hard, local cultivars, 'Huntsman' and 'Cockpit'.

Every ship's company on a long voyage would soon have learned the antiscorbutic character of apples, and the apple was practically the only fruit until the lemon to travel this way. (Hard lemons, sometimes limes, were commonly carried on 19th century British navy ships, hence the American term for a British sailor: limey.) From the robust and tough-skinned phase 3 apples that may also have bruised the heads of Alexander the Great's crews were selected the specimens that, packed in barrels of sand, bran, or sawdust, would have survived the motion of a ship and crossed the oceans with the first colonists. It is likely that every fighting ship, beginning in the 16th century, carried barrels of sand, bran, or sawdust to render the decks less slippery when the action was fierce.

Colonies

Notwithstanding the development and dissemination of the techniques of grafting (Chapter 4), the habit of collecting a seedling pippin was widespread and common from early on. The hedgerows and coppices of most of western Europe must have been scattered with feral apples of highly variable quality. But here and there, hedge-laying and ditching teams, particularly, must have come across outstanding, large, sweet seedlings in the autumn after the harvest, as authors from Rabelais ("make seminaries [seedbeds] with their pippins in your country," from *Gargantua and Pantagruel*) to Shakespeare (Chapter 5) testify.

As the first colonists moved across the ocean to the New World, it would therefore have seemed perfectly natural to save every seed from the ship's kitchen waste, clean it, and plant it in due time. The fierce winters of the North American continent would have done the rest. Apple seeds derived from European stocks were almost certainly germinating in the plantations of the colonists—French, Dutch, and British—along the eastern seaboard of North America in the first years of the 17th century (Bazeley 1991). In spite of the fact that the number of individual genotypes transported across the ocean must have been very small, each colonization seems to have resulted in an explosion of new cultivars.

Apples, probably in the form of seeds in the first instance, and selections in the form of scion wood rather later, were soon established in the Tidewater region, the area of eastern Virginia penetrated by the James, York, and Potomac Rivers and thus susceptible to tides in Chesapeake Bay. Samuel de Champlain (he of Lake Champlain on the northern border of Vermont and New York, and adjoining Québec) had planted the first Normandy apple trees on the heights above the city of Québec, and by 1626 the physician on his expedition, Louis Hébert, had nearby a whole orchard of apple trees (Martin 2000). Apple seeds, along with quince kernels, were listed in the *Memorandum* of 1629 of seeds to be sent from England to the Massachusetts Company.

The vigorously germinating apple seeds in America, with its sustained low winter temperatures, gave rise to a distinctive new spontaneous cluster of cultivars such as 'Jonathan', 'Wagener', 'Red Delicious', and 'Golden Delicious', successful in that more extreme climate. 'Golden Delicious' was not, as commonly supposed, either French or an act of deliberate hybridization; it was found as a chance seedling of unknown parentage in a hedgerow in Clay County, West Virginia, in about 1890. The original tree survived in a padlocked steel cage with a burglar alarm until the 1950s (Pollan 2001). 'Golden Delicious' achieved fame, and some notoriety, by being brought over to Europe after 1945 by the U.S. Department of Agriculture as part of the Marshall Plan for the restoration of European agriculture. 'Golden Delicious' is now the most widely grown commercial cultivar in France, and it (or 'Red Delicious') is possibly the most commonly grown apple in the world, though 'Granny Smith' is gaining ground.

The speed with which variants of the flexible apple genome were selected can be inferred from a list of cultivars written in the third quarter of the 18th century and now preserved at Colonial Williamsburg, Virginia. Both to the East and to the West, through colonization and in the reverse direction, individual cultivars to and from the British Isles have rarely been successful (Darwin 1868), but there would obviously have been some social cachet in growing "old English" varieties in a colonial setting: 'Baker's Nonsuch', 'Baker's Pearmain', 'Bray's White Apple', 'Clark's Pearmain', 'Dutch Pippin', 'Ellis', 'Father Abraham', 'French Pippin', 'Gillese's Cyder', 'Golden Pippin', 'Golden Russet', 'Green Old England', 'Harrison's Red', 'Harvey's Holland Pippin', 'Horse', 'Hugh's Crab', 'Lightfoot's Father Abraham', 'Lightfoot's Hughes', 'Lone's Pearmain', 'Ludwell Seedling', 'May', 'Newtown Pippin', 'New York Yellow', 'Non-Pareil', 'Old England', 'Pamunkey Eppes', 'Red', 'Royal Pearmain', 'Ruffin Large Cheese', 'Sally Gray's', 'Sorsby's Father Abraham', 'Sorsby's Hughes', 'Summer Codling', 'Westbrooke's Sammons', and 'Winter Codling'. Of these, in spite of the manifestly English resonance of almost all their names, only 'Golden Pippin', 'Golden Russet', and 'Non-Pareil' may actually be the names of true English cultivars.

Philip Forsline (pers. comm. 2003) reports that all the native North American apples are strikingly astringent. The native Americans, who in some areas had developed a very sophisticated horticulture with crops such as beans, squash, and corn (maize), do not seem to have exploited the potential of the native crabs such as *Malus angustifolia, M. coronaria, M. fusca,* and *M. ioensis,* though *M. coronaria* seems occasionally to have been used for cider (Hedrick 1919). In the modern period, besides use in a number of medicinal preparations, introduced and native apples are certainly eaten fresh or dried, and made into jelly. The Kitasoo of coastal British Columbia store apples in water sealed with mammal or fish grease or oil for winter use (Moerman 1998).

Grafting was not very widely practiced in the Americas in the 17th and 18th centuries, or even into the early 19th, and the settlers relied upon the fecund germination of apple seed, as do the peoples of Inner and Central Asia still. Dutch, English, and French cultivars were certainly taken across, from time to time, in the form of whole grafted plants, as they were elsewhere. William Coxe (1817) noted that in Philadelphia he grew several of the 'Calville' cultivars from France, 'Leathercoat [sic] Russet' and "Ribstone Pippin" along with 'Redstreak' for cider. Whole, grafted, transported apples sometimes behaved in unpredictable ways, however. As Darwin (1868) noted, 'Ribston Pippin' changed into a fastigiate form when transplanted to India (Chapter 5).

But on the whole, European apple cultivars were generally not satisfactory in the American climate. They did, however, often set seed, where they were fertile and could find compatible partners. The seeds would have germinated more readily after the extreme continental winters, far more so than was generally the case in the soft winters of the "Old Country" or the rest of maritime Europe. The European cultivars may have hybridized to some extent with the

local native crabs, but although these hybrids have some importance in the raising of ornamental *Malus* cultivars, they do not appear to have entered significantly into *M. pumila* lineages. In fact, *M. pumila* rarely acts as a viable pollen
source in hybridizing with native American crabs (Dickson et al. 1991).

The genetic diversity that accumulated in apples as a result of the spectacular
germination after the cold winters soon became considerably greater than that in
their European counterparts. The potential for crosses between cultivars was enormous, such that eastern North America from Virginia northward to well over the
Canadian border became a vast experimental station that served to screen huge
numbers of open-pollinated seedling apple cultivars, mostly unwittingly.

Jan van Riebeeck is reported to have taken sweet apple seedlings to Cape
Town, South Africa, which town he founded in 1652 (Hedrick 1919). Deciduous
fruits are now cultivated beyond the Hottentots Holland Mountains, where the
elevation is higher, about 350 m (1,150 feet), and the average winter temperatures
are considerably lower than in the first settlements around Cape Town—occasional frosts occur. Similarly, 'Granny Smith' arose as a chance seedling near
Sydney, Australia, in the 1850s (Mabberley 2001). There, seeds would have enjoyed no cold-chill, and it is possible that the Tasmanian apples from which the
seeds came had suffered frost in that island's much cooler climate before arriving
in Sydney. Today, Tasmania is known as "The Apple Isle."

However, what has certainly happened in the era of the apple's colonial
expansion is that unconscious selection for low-chill apple cultivars has taken
place. Low-chill genomes are now a routine part of apple breeding programs in
an attempt to accelerate the acceptance and spread of new cultivars. On the other
hand, in regions with intense continental climates, as in the American Midwest
and much of Russia, extended cold-chill cultivars are sought.

CHAPTER 7

Beyond dessert:
Cyder and ornamentals

APPLES ARE RICH IN FLAVONOIDS, which counter free radicals, and vitamin C, which survives the making of cyder and cider in which, because of fermentation, vitamin B_{12} is also found. Cider is not typical of Asia but was known to the western Classical civilizations and is characteristic of northwestern Europe, whence it was taken to North America. Seeds from waste generated in its production there were spread over the United States by Johnny Appleseed and others. In Asia and elsewhere, drying apples for food is a very significant industry.

Many important modern cultivars are intraspecific hybrids, though their presumed parentages are often challenged by modern molecular techniques. A very small number of interspecific hybrids can be counted among the merely ornamental crab apples.

THE APPLE in its various forms and manners of presentation is an excellent, but not outstanding, dietary component. Fresh apples, though varying from cultivar to cultivar, are a source of vitamin C, potassium salts, carotenoids, and dietary fiber (McWhirter and Clasen 1996). The dessert apple 'Ribston Pippin' (Plate 8) contains 30 mg of vitamin C for every 100 g of flesh, and some of the cooking apples, particularly triploids such as 'Bramley's Seedling', come close to that figure. Although understood in an empirical way for at least 400 years, when the antiscorbutic factor and the chemical nature of vitamin C became scientific knowledge (Hall and Crane 1933), tables of relative nutritional values were published, often in a laudable bid to increase consumption.

Apples are also particularly rich in a class of compounds termed flavonoids. The chief effect of these compounds on human health may be to counter the disease-triggering entities called free radicals. These are destructive molecules that may be initiating factors in conditions that give rise to chronic conditions such as heart disease, cancer, diabetes, and asthma. Apples appear to be significant in supplying radical-quenching compounds. Specific flavonoids in apples are cate-chin and quercetin, both of which appear to be potent in reducing radicals.

When converted into cyder or hard cider, the juice, unpasteurized, retains not only most of the vitamin C but also a generous allowance of vitamin B_{12} as a result of the fermentation. Since there is no ordinary vegetable source for it, this vitamin B_{12}, from the fermenting yeast, would have been valuable, if not actually lifesaving, in medieval, meat-starved, rural Britain or France, or colonial America, in the middle of winter. Pasteurization, an essential part of mass-marketing, has virtually removed this vitamin from almost every commercially produced apple juice.

Cyder and cider

Cyder is a word that occurs in almost all the Indo-European languages, but it may have existed longest in those languages in the warmer regions. The modern English term *cider*, spelled almost universally *cyder* until the 19th century, evolved from the Middle English *sidre*, which itself came from an Old French word of the same spelling. That word is probably cognate with late Latin *sicera* (Greek *sikera*).

Taking a cue from that comprehensive text, *The History and Virtues of Cyder* by the late Roger French (1982), *cyder* should be clearly distinguished from *cider*. It appears that with production beginning in England shortly after the Conquest, when it was made solely and directly from the pressed apple juice with no added water, cyder in his definition had a an alcoholic composition approaching wine strength. Either legally or illegally, the cyder could then be diluted with considerable quantities of water of whatever potability to become cider, and the ethanol content, combined with the tannins from the apples, would still effectively destroy any bacteria present.

Cyder or cider apples are traditionally divided into four categories: sweets, bittersweets, bittersharps, and sharps (Table 2). The third and fourth categories are so astringent as to be inedible to conventional taste. Bittersweets contain more than 0.2% (weight to volume) of tannins and less than 0.45% acidity, due principally to malic acid. Bittersharps have the same high level of tannins but an acid content greater than 0.45%. Most cider makers add a percentage of conventional sweets (dessert apples and cooking apples, such as 'Golden Delicious' and 'Bramley's Seedling', respectively) to the fermentation.

There does not seem to be a tradition of cyder making in the lands of Inner and Central Asia, though apples along with other fruits, usually when dried, were and still are often steeped in boiling water to make a range of infusions.

TABLE 2. Cyder and cider apples*

CULTIVAR	SEASON	TYPE	MALIC ACID (%)	TANNIN (%)
'Bulmer's Foxwhelp'	early	bittersharp	1.91	0.22
'Somerset Redstreak'	early to mid	bittersweet	0.19	0.28
'Sweet Alford'	mid	sweet	0.22	0.15
'Sweet Coppin'	mid	sweet	0.20	0.14
'Ashton Brown Jersey'	late	bittersweet	0.14	0.23
'Bramley's Seedling'	late	sharp	0.85	0.05
'Chisel Jersey'	late	bittersweet	0.22	0.40
'Dabinett'	late	bittersweet	0.18	0.29
'Golden Delicious'	late	mildly sharp	0.45	0.06
'Kingston Mill'	late	bittersharp	0.58	0.19
'Médaille d'Or'	late	bittersweet	0.27	0.64
'Tom Putt'	late	sharp	0.68	0.14
'Yarlington Mill'	late	bittersweet	0.22	0.32

*Different methods of analysis may give different values for chemical constituents; the figures here should be treated only as comparative.

It may be that the heavy equipment, the movement of liquids in hard bottles or casks, and complex, essentially immovable, crushing mills were ill suited to what had been for many years mobile, semi-nomadic populations. There seems little doubt that the development of the grinding and crushing of distinct cultivars of fruit to produce fermented liquors emerged in parallel with, though probably later than and often interwoven with, the techniques used for olives, grapes, and to a certain extent those for ale and beer making.

There also seems no doubt that the invention of cyder (and perry) provided an immensely powerful but again unintentional engine for the evolution of the apple (and pear). The quantity and diversity of seeds resulting from cyder apple processing must exceed those from the deliberate, controlled cultivation of a small number of dessert and cooking apple cultivars.

Cyder making, on the other hand, particularly in rural communities relying on random apple seed germination, tends to bring together a more diverse gene pool than any other system of cultivation. Thus, as the apple was moved west out of Inner and Central Asia into more settled societies, not only did it bring vital nutrition to rural communities in winter, but it also provided potable liquor rich in vitamins. At the same time, its own evolution was accelerated. These parallel developments were particularly important in North America.

The grinding and crushing of apples has always been carefully managed, whether in small-scale or industrial processing, so that the tissue is reduced to small crumbs but emphatically not a mush, and the seeds themselves are not broken. Seeds would tend to add a taint to the brew if broken because they contain amygdalin (vitamin B_{17}), which releases cyanide when it comes into contact with water. Thus the very important step of separating the seed from the placental tissue, which will ensure successful germination, is achieved mechanically in the apple mill or grinder. The jaws and guts of the bear and horse will also separate the seed effectively and generally not damage it, whereas the human jaw or bird beaks are not very effective separation mechanisms. Cooking a whole fruit is also obviously disastrous to the future of the apple seeds. Given the usual human way of eating an apple, even the seeds of a dessert apple are unlikely to find fertile ground. Moreover, the narrow selection of apples grown in orchards tends to restrict the genetic range.

For any reasonable-scale cyder (or perry) production, the most precise crushing or mincing of the fruit pulp is necessary. The apple flesh needs to be reduced to a size not much smaller than that of a child's fingernail, but not to a purée from which no free juice can easily be extracted. The cyanide-containing seeds must not be crushed. It would seem from the investigative work of French (1982) that

"Ingenio" from John Worlidge's *Vinetum Britannicum* (1676).

the necessary machinery has a direct evolutionary connection with the apparatus designed for the crushing of olives and the extraction of their oil. Such a history might suggest, as French does, that the particularly tannin-rich cyder apples, too, might have had a more southern origin than the northern dessert or cooking apple. But this speculation remains to be tested by modern molecular methods.

Considerable inventiveness was applied to the machinery required. By the time of John Worlidge (1676), with his "ingenio" (the term also applied to the sugarcane-crushing mills developed in the Caribbean at almost the same time), a precise expertise was brought to the milling process. In addition, the ingenious Worlidge sought to remedy the persistent defect of all screw presses since the earliest times, namely, that the pressure is discontinuous. Whether his remarkable machine was ever built must remain open to doubt, but it can be seen from the published drawings that it was certainly capable of maintaining a steady pressure on the mass of carefully minced apple flesh.

Worlidge also turned his attention to the various stages of the fermentation process, bottling, aided and abetted by the rapid technical advances in glassmaking in Britain, and storage of real cyder (an accessible account of much of Worlidge's work can be found in Juniper and Juniper 2003). His barrel, more correctly a stund or stound, is carefully designed so that as the contents are drawn

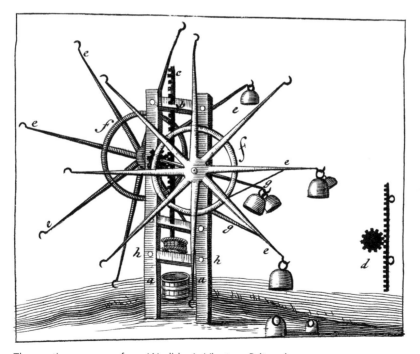

The continuous press from Worlidge's *Vinetum Britannicum*.

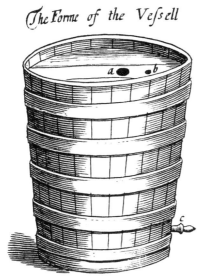

The Forme of the Vessell

A stund or stand or stound from Worlidge's *Vinetum Britannicum*: a, the bung hole; b, a small vent hole; c, the tap.

down through the tap, the natural crust that forms on the fermentation does not break but is consolidated. In the standard barrel of the period, the crust would have stretched and broken, permitting bacterial or undesirable yeast contamination.

Other authorities of the period were obviously keeping their eyes open, such as the anonymous "Lover of Planting" (1685), noting trees that "will of themselves bear good Cyder-Fruit, (which he may guess at by the broadness and largeness of the Leaves they bring forth)," suggesting that extensive genetic segregation of characteristics was already present in the semi-wild or feral populations of *Malus pumila* in British hedgerows, and that cyder makers were aware of the potential offered by such variety.

Early Europe

It seems possible that the technology, the cultivation of the special fruit required (high in tannin), and the tradition of drinking cyder came into western Christendom from the non-Muslims of Moorish Spain. From France it seems likely that the habit came to England after the Norman Conquest. Much earlier, the Greeks and Romans doubtless made cyder. Before even that, it is possible that some of Alexander the Great's heroic, occasionally lethal, drinking bouts (Lane-Fox 1973) may have involved cyder.

In southern France a beautiful Roman mosaic dating from the first half of the 3rd century records the whole orchard industry, including cyder making (Plate 34). There is no reason to suppose that such a valuable and widely disseminated technology would have disappeared during the Dark Ages, not least because of the investment in the robust, long-lasting crushing and pressing equipment. It is likely, too, that Charlemagne's sophisticated horticulture preserved these important secrets. Cyder making, at least in Europe through the Dark Ages, seems to have been sustained only in southern and southwestern Britain, in Brittany and Normandy (ancient Armorica) in France, and in the region of

Asturias in northern Spain (Pollard and Beech 1957). But it should not be forgetten that there are long-standing cyder-making traditions in, for example, parts of modern Turkey, which might be consistent with contact with the Berbers and Moorish Spain.

Britain

The first solid documentary evidence for cyder making in Britain comes from near Norfolk during the reign of John, king of England 1199–1216. Later, Geoffrey Chaucer (about 1342–1400) definitely knew about cyder because in *The Canterbury Tales* he has some of his pilgrims drinking it. There is a drink with which the ancestors of the British were wont to regale themselves: it was called lambswool, a word derived from the Middle Irish *la mas abhal* (Gaelic *ubhail*) in which is the root word *lammas*, which signifies the day of the apple fruit (Hogg 1851). The drink was composed of ale and the pulp of roasted apples, with sugar and spices (Gerard 1633).

Cyder cultivars, with their peculiar properties, were best grown in the counties of the west of England, namely, Herefordshire, Worcestershire, Gloucestershire, Somerset, Devon, and Cornwall. Ralph Austen (1657) wrote,

> I need not tell *Herefordshire,* and *Worcestershire* men, the good properties of *Perry* and *Cider,* they know by experience it is *Alimentall and Physicall,* that it is proftable not only for *Health,* but also for *long life,* and that *Wines* made of the best kind of *Apples and Peares,* is a speciall *Cordiall,* chearing and reviving the spirits making the heart glad as *wine of Grapes.* And it has been observ'd that those who drinke *Cider* and *Perry* daily, or frequently as their common drinke, are generally *healthy persons* and *long-lived.* . . . And that it will begger a *Physitian* to live where *Cider,* and *Perry* are of generall use.

Edward Drope, writing a preface to his late brother Francis's *A Short and Sure Guid in the Practice of Raising and Ordering of Fruit-Trees* (Drope 1672), declared, "men Living to great Ages in the Cider-Countryes, both Active and strong, as may appear by a storie, I shall here insert, which I borrowed out of an Honourable and Learned Author [Lord Verulam] that at a Wake in Herefordshire, a Daunce was performed by eight men whose ages, added together, amounted to eight hundred years."

Scudamore's 'Redstreak'

Shortly after the assassination of the duke of Buckingham in 1628, Viscount Scudamore (1601–1671) retired to his estate of Holme Lacey in Herefordshire under a political cloud, the result of his support for Buckingham. There he concentrated on planting orchards, making cyder in an attempt to match the much superior French product of the time, and raising seedling fruit trees. Among the most spectacular of his raisings was 'Redstreak', which became the byword for quality cyder and cider. It is reported that in 1639, when he returned from being ambassador to Louis XIII of France, the viscount's Christmas festivities consumed some 1,200 British gallons (5,500 liters) of cyder (Ward 1988, 1992). Ralph Austen (1657), who built Oxford's first cyder factory, was the first to mention 'Redstreak' in print, and John Evelyn (1664) in his *Pomona; . . . Concerning Fruit-Trees in Relation to Cider: The Making, and Several Ways of Ordering It*, noted that by Scudamore's example, "all Hereford-shire is become, in a manner, but one intire orchard . . . The *Red-strake, Bromesbury-Crab*, and that other much celebrated *Wilding* call'd the *Oken-pin*, as the best for *Cider;* though for sufficient reasons none of them is comparable to the *Red-strake*." The same year Beale wrote, "Yet the choice of the *Graff* or *Fruit* hath so much of prevalency, that the *Red-strake-Cider* will every where excel common *Cider*, as the *Grape* of *Frontignac, Canary*, or *Bacharach*, excels the common *French Grape*."

Vinetum Britannicum

As cyder making began to be described in the literature, the texts culminated in 1676 in the brilliant, comprehensive, and lucid *Vinetum Britannicum* by John Worlidge. Even at this early date, Worlidge was obviously familiar not only with 'Redstreak' and all its virtues but with practically all other liquors from every known distilled beverage, including rice wine as well as coffee, tea, and coconut milk. He seems also to have been an ingenious engineer, proposing mechanical improvements for the conversion of the apple into cyder and the modern style of bottling of the liquor.

Throughout the rest of the 17th century (Haines 1684), the whole of the 18th (Philips 1708, so popular that it went through no fewer than seven editions up to 1791, and Stafford 1755), and into the early 19th, books on cyder poured from the presses (Janson 1996), stimulated by the fact that the ongoing wars with France had raised the price of wine to outrageous levels. The 1803 edition of

William Forsyth's *Treatise* lists and depicts 197 cultivars of apples, and many are those for cyder.

Adulation of the cyder apple reached a pictorial peak with the publication of the exquisite *Pomona Herefordensis* of Thomas Knight in 1811. 'Redstreak', among others, was beautifully illustrated by Alice Ellis and Edith Bull in Hogg and Bull's (1876–1885) *Herefordshire Pomona*. But by the end of the 19th century and into the 20th, in part due to the fact that a sustained peace with France had broken out and the price of good wine fallen, the cyder industry in Britain went into decline.

Cyder and cider making in the United Kingdom has suffered much turbulence in supply and demand. Cyder and cider making are undergoing a major revival; for example, more than 100 million British gallons (450 million liters) of cider were made in Britain in 1997. The industry in Britain has also enjoyed a literary revival comparable to that of the 17th century (Copas 2001). The apple harvest in the Tamar Valley of Devon and Cornwall has received particular attention (Spiers 1996). The industry was sustained and revived from its moribund state by the Long Ashton Research Station near Bristol, from 1903 (Marsh 1983). Quite apart from the advanced study of soil conditions, rootstocks for grafting, storage, and preparation (Williams 1987), the identification of proper cultivars, very important for a sophisticated industrial operation, has been studied in depth (Williams and Child 1965).

France

It has to be admitted that in literary terms (as well as liquid output), the Anglophones cannot approach the sophistication of *Pommiers à Cidre* (Boré and Fleckinger 1997). The industry in France is ancient, large, and diverse, encompassing not only many cyders and ciders but also apple juices and many varieties of the distilled eau-de-vie called calvados. Visitors to Normandy and Brittany may still be asked if they would wish to add *un p'tit calva* if their coffee is not strong enough. The distilling of cyder does not appear to have taken serious root in any other country except to a small extent in southwestern England, but "Compared with the cider apple orchards in England, French orchards have preserved a much greater of range of cultivars that can be used for the making of apple juice, cider, and calvados" (translated from Boré and Fleckinger 1997). The individual right to distill cyder legally and make one's own calvados (*Les Bouilleurs de Cru*) was fiercely defended by families and the owners of particular orchards, such that the practice survives to a limited extent. However, notices on the farm gate saying *Calva de Ferme*, calvados of the farm, are now a thing of the past.

The French industry got off to an early start. The establishment of orchards and large-scale production of cyder were described in accurate detail by Julien Le Paulmier de Grentemesmil in *De Vino et Pomaceo Libri Duo* in 1588 (see Janson 1996). By 1589 it is recorded that there were already 65 cultivars of cyder apples being grown in Normandy alone; by the end of the 19th century there were 300. Now the industry seems to be centered in Brittany and Normandy, with few orchards south of Paris. Many hundreds of cyder cultivars are currently listed (Boré and Fleckinger 1997), but most cultivar names are extremely local—some carry what are obviously very ancient Breton (Celtic?) names such as 'Chuero Bruz' and 'Chuero Ru Bihan', names meaningless to the average French person. Indeed, they must be almost as difficult to pronounce for a native Francophone as they are for an Anglophone. Is it possible that some of these names have persisted from before the Roman invasion?

Unlike dessert and cooking apples, and numerous pears and plums, very few French cyder cultivars are found in lists of English apples. And despite the early French connection, none is in use in North America as far as known. This lack of overlap has led to suggestions that French cyder cultivars arose as the result of hybridizations between the Roman cultivars of the sweet *Malus pumila* as they spread up through northwestern Europe and the local, tannin-rich *M. sylvestris* (Plate 3). This speculation remains to be tested. Such sharp apples would have been selected as valuable for a local industry, but because of their relatively low value in relation to their bulk, they would not be transported far from their original orchards. Again, unlike their very sweet or culinarily valuable large cooker equivalents, they did not become known to a wider public. Americans and Canadians, at least in the early days of colonization, consumed large quantities of cyder, cider, and applejack but with few exceptions do not seem to have developed their own cyder cultivars.

North America

Hard (cyder) and soft cider were made in the American colonies from an early date. In 1647 it was recorded that "twenty butts of cyder" were made in Virginia, and by the mid-1700s both cider and cider vinegar (essential for the storage of many foodstuffs in tropical conditions) were being exported from Virginia to the West Indies (Calhoun 1995). Cyder probably helped, not only from the nutritional point of view, on some of these long sea voyages.

Evidence for the importance of the apple in what were still pioneer communities is exposed at a colonial archaeological site. In Boston, Massachusetts, a privy located at Cross Street Back Lot reveals a wealth of detail concerning the diet of North American colonists from the late 17th century. Some of these plant remains, principally the small seeds, were of fecal origin, and among these there were abundant seeds from apples, pears, blackberries, elderberries, squashes and pumpkins, and a range of spices (Dudek et al. 1998). The presence of fruit-tree seeds suggests that orchards were well established by the mid-17th century. At that time, apples and pears were much used for hard ciders as well as for fruit pies and other dishes.

A true cider industry had almost ceased to exist in the United States by the end of the 19th century. Such that was there was made from rejected dessert apples with no respect as to quality. The temperance movement and Prohibition almost brought about the demise of cider, a state from which it is recovering.

Johnny Appleseed

John Chapman was born in Leominster, Massachusetts, in 1774 and died in Fort Wayne, Indiana, in 1845. *Chapman,* which is Old English for an itinerant salesman or peddler, might give us a clue as to his robust traveling and retailing skills. Nicknamed Johnny Appleseed, he spread, along with his Swedenborgian doctrine, the random hybridizations of a near infinite number of apple pollination events across much of a continent. He seems to have been more successful with seed dispersal and orchard building than with religious conversion. Chapman spread seeds, mostly waste from cider mills, from farm to farm over virtually the whole of what was then the western frontier, through the Ohio Valley well up into what was called the Northwest Territory and over what is now the Canadian border (Martin 2000). One of his favorite cultivars is said to have been 'Rambo', which originated in Sweden. It is reported that sometimes he would completely fill one hull of a double-hulled canoe with apple seeds. He spurned grafting and was quoted as saying, "They can improve the apple in that way, but that is only a device of man, and it is wicked to cut up trees in that way. The correct method is to select good seeds and plant them in good ground and only God can improve the apple" (Pollan 2001).

There was another reason for the spread of apples. It is not often that the law directly aids evolution, but a land grant in the old Northwest Territory in the last quarter of the 18th century required a settler to plant at least 50 apple or pear trees as a condition for the granting of the deed. The aim was to suppress real estate speculation by encouraging homesteaders, literally as well as metaphorically, to put down roots. Since a typical apple tree takes at least 10 years to fruit from seed, an orchard indicated an enduring settlement. It was from these random apple forests that, later, tree grafters selected their choicest specimens (Pollan 2001). John Chapman would not have been amused.

Thus, as a result of human dispersers of whom the best known was Johnny Appleseed, a substantial portion of North America became a multiedaphic, multi-climatic, north–south, seed-testing station, much as the great east–west trade routes through and around the Tian Shan at a much earlier date were an elongated, unintentional seedling selection ground.

The vast majority of Chapman's trees would have been scarcely edible, bitter specimens, the "spitters" of local dialect. But most apples grown in North America, from the earliest settlements almost to 1900, were probably drunk, not eaten (Pollan 2001). Hard and soft cider were beverages both of urban and rural communities. Bitterness was often advantageous in cider making (Table 2). Johnny Appleseed brought alcohol to the frontier, spread apple seeds, and unwittingly drove forward the evolution of the apple.

Virtues of cyder

Although it is very likely that cyder making was already present in England, at least on a small scale, it spread throughout Britain after the Conquest. The reasons for this popularity are not difficult to find. Mild alcoholic drinks, where pure water was in short supply, particularly in populated areas and even more so in the depths of winter, provided not only safety but also the benefit of vitamins C and B$_{12}$. Diluted with water, during or after the apple's fermentation, the resulting low-alcohol liquid, often termed cyderkin or small cider, was considered a safe and practically nonintoxicating drink, even for children. Cider or small cider was not to be supplanted, certainly for rural workers, by any other form of liquid, until tea, with its double safety advantage of boiled water and rich tannins, became readily available in the 19th century. Cider was part of the daily wage of rural workers in Somerset even up to World War II (Copas 2001).

In western Europe the equivalent drink was beer or ale, with small beer, the last, weak fermentation, for children. Small beer was considered a safe drink for the children at least until the end of the 19th century. It was served at midday and in the evening at the tables of Oxford and Cambridge colleges as late as the 1920s because local water supplies were so unreliable.

On the American frontier, the grains necessary for beer or ale could scarcely be spared. The grain that could be grown was converted not into beer but distilled to whiskey, which as a high-value, low-volume product was often taken to market for sale. Apples, unlike the grains, would grow on almost any rough patch of land or rocky slope and required minimal attention. Apples were infinitely more tractable than labor-intensive vines, even where the climate was suitable for grapes. Moreover, in a hard season, when fodder was in short supply, as in 1815–1816 (Coxe 1817), pigs and horses could be kept fat on the poorer grades of apples until late December, and cattle in the fields were fed on the pomace, the waste of crushed pulp and seeds from the cider mills (in some texts it is written *pummice* and is the same as *murc* or *pouz* of earlier writings).

The pips that Johnny Appleseed transported came from the mounds of pomace left over in the autumn after the presses that stood outside every farmer's barn had done their work. Diligent as he was, John Chapman could never have collected the amounts of seed he used by splitting open apples and prizing out the pips. It is likely that he washed the mass of pomace through a riddle or coarse sieve. Doing so would have further scarified the seeds and separated out both the useless pulp and the small, nonviable pips. There is written evidence of this general practice (Coxe 1817 in his Chapter 2, "On the Management of a Fruit Nursery"): "The seeds generally used for this purpose, are obtained from the pomace of cider apples—they may be sown in autumn on rich ground, properly prepared by cultivation, and by the destruction of weeds, either in broad cast, or in rows, and covered with fine earth; or they may be separated from the pomace, cleaned and dried, and preserved in a tight box or cask to be sowed in the spring: the latter mode may be adopted when nurseries are to be established in new or distant situations, the former is more easy and most generally practised."

In 1810, 198,000 barrels (about 24 million liters) of cider were made in Essex County, New Jersey, alone. Unlike Britain and France, however, very few American apples were listed specifically for cider making. 'Hewe's [Hugh's] Crab', which may have originated in Virginia in the early 18th century, is a rare exception. Another to be considered is 'Smith's Cider', which probably arose as a seedling in Pennsylvania about 1776. But even this latter apple is not exclusively used for cider and is best considered more of a dual-purpose cooker. Defying the general rule that English apples do not thrive in the New World, the famous 'Redstreak', though probably no longer the original, was widely grown as a fine enrichment to the cider press, particularly in the southern United States: "it ['Redstreak'] has been cultivated extensively in this country, by the descendants of the English settlers in New-York, New-Jersey, and Pennsylvania. The climate of America is supposed to have revived the character of this apple, which had deteriorated in its native soil from the long duration of the variety" (Coxe 1817).

If allowed to ferment for a few weeks with natural yeasts, cyder will reach an alcoholic strength about half that of wine. This fermented juice is called hard cider in the United States. Distillation of fermented apple juice (calvados) was not common in North America because it was scarcely necessary. A barrel of hard cider, if placed out on the porch and exposed to the depths of a good North American winter, could be transformed. A natural phase separation would take place, and some of the water would turn to pure ice on the surface. This ice could be thrown away and the process repeated until the residual fluid, applejack, reached a level of 66 proof, or 33% alcohol. In his chapter, "Of the Concentration of Cider by Frost," Coxe (1817) gave a full description of the technique. The advantage of the freezing method was that there was no need for distillation equipment, which might provide evidence of illicit activity and attract the attention of the tax gatherer.

Naturally fermented cider not only retains a high level of vitamin C from the original fruit, but also other vitamins. But during pasteurization these essential vitamins, being mostly heat sensitive, are almost completely destroyed and in many modern, mass-marketed products are returned artificially. Cyder made in England in the 1970s in the way it was in the 17th and 18th centuries, using a popular modern bittersweet apple such as 'Strawberry Norman', yielded 33.8 mg vitamin C per 100 g (French 1982). In other drinks to which water is added, and which should be termed cider, the vitamin C content, as well as the ethanol, will be proportionately diluted. A commercial, pasteurized cider might return a figure of only 3.3 mg vitamin C per 100 g.

Other methods of apple preservation

Apples not only may improve in quality if they are kept whole, but there is a range of processing methods, besides the preparation of cider, that preserve their excellent nutritional qualities. The names of old apple cultivars such as 'Douzins' (a corruption of the French *deux ans,* 2 years—the name is preserved in the 18th century cooking apple 'Hambledon Deux Ans', famed for its keeping qualities) may indicate their storage capability. 'French Crab' (also called 'John Apple'), apparently the parent of 'Granny Smith' (Mabberley 2001), was also reputed to keep 2 years. Many apples can be stored whole if protected from severe frosts, clamped (buried in straw-lined pits in earth below the frost depth), or sliced and dried.

As the ancient Sumerian grave of Queen Pu-abi has shown, drying apples is a very old method. Not only does this technique require little immediate or long-term energy expenditure, but it both preserves the full food value of the fruit and renders otherwise tannin-rich, bitter fruits edible (Wiltshire 1995, Renard et al. 2001). It seems likely that the soluble tannins of the fresh fruit are, by the action of cutting and drying, released from the cell vacuoles and bound irreversibly with proteins and carbohydrates from the cytoplasm. The importance of the method can be gauged from the fact that in the United States, where the storage of whole, fresh apples through a continental winter was possible but tricky and the failure rate significant, one shipper alone in Baltimore sent 2,000 pounds (900 kg) of dried apples to Germany in 1876 (Calhoun 1995).

There was yet another method for the overwintering conservation of apples in North America: the widespread production of apple butter. A mass of apple pulp, often using damaged fruit, was cored, cooked, sieved, and dehydrated by boiling to a thick butter, sometimes with the addition of spices such as cinnamon or cloves. This culinary procedure was one of the few that would actually release intact apple seeds for possible future germination. Almost any apple would do, but connoisseurs favored the cultivars 'Wolf' and 'Buff' (Calhoun 1995). Sometimes, too, a little cider was added. As much as 12 hours of cooking might be required, and the resulting viscous, almost solid butter was preserved in crocks covered with thick paper and stored in cellars.

Jia Sixie's (1982) *Qi Min Yao Shu*, necessary skills for the masses, records recipes for preserving apples from 6th century China:

> To make *naichao* [apple flour]. Collect overripe *nai* [*Malus pumila*], put them into an earthenware pot, and cover the mouth with a bowl to keep out flies. After 6 or 7 days, when they are very rotten, pour in enough spirits to cover them and stir vigorously until they become like rice gruel. Add water, stir again, then strain to remove the skin and pips. Leave for a good while to settle, then drain off the liquid, again add water, and stir as before. Stop when there is no more bad smell. Drain off the liquid, cover with a cloth, and use ashes to soak up the juice, as if making rice flour. When there is no more juice, cut up into pieces as big as the back of a comb [?] and dry in the sun. Finally, grind to powder. Sweetness and sourness are well balanced, and the fragrance is exceptional.
>
> To make *linqinchao*. When *linqin* [possibly *Malus asiatica*] are red ripe, cut them up and remove the pips, cores, and stalks. Lay them in the sun to dry. Grind them or pound them and sieve through fine silk pongee [filter cloth]. Grind or pound the coarse fragments again until as fine as possible. Take a spoon 2.5 cm (1 inch) square full of this powder and put into a bowl of water. It makes an excellent broth. If the stalks are not removed, then it is very bitter; if the seeds are left in, it will not keep through the

summer; if the cores are not discarded, then it is very sour. If it is eaten dry, mix one measure of *linqinchao* with two measures of rice *chao* [cooked grain ground into flour, rather like Tibetan *tsamba*]. The taste is really good.

To make *naifu* [dried apples]. When *nai* are ripe, cut them through the middle, dry them in the sun, and they are ready.

Still today, late August through September, it appears that almost every community in central and eastern Europe and Inner and Central Asia is in the business of drying fruit. Particularly in Asia, where the heat is fierce until late September, it seems as if every corrugated iron roof, concrete surface, and even stretch of smooth, black tarmac is commandeered for fruit drying. Grapes and apricots are the major crops, principally for export, but apples are not far behind. Any problems with the traffic attempting to reclaim a useful length of hot, smooth highway are thwarted by placing a small, immobile member of the family at each end of the drying stretch (Plate 30).

Apple growing

Worldwide production of apples now runs into the many millions of tons a year. The United States has some 7,500 commercial growers in 36 states (Hanson 2005). The state of Washington has been the largest producer (particularly of 'Red Delicious') since the 1920s, in 2004 growing more than half the country's production of 4.7 billion kilograms (10.4 billion pounds; Choy Leng Yeong 2005). As the state's most valuable agricultural product, the apple has become so iconic that there are even Golden Apple television awards, and the most significant interuniversity American football match in Washington is the Apple Cup. By 2003, Americans on average each ate more than 7 kg (more than 16 pounds) of apples a year, and though that is 20% less than in 1989, it is still second only to bananas in terms of fruit consumption. The United States now exports nearly 500 million kilograms (1.1 billion pounds) of apples a year, but this effort has been overtaken by China's export production of more than 850 million kilograms (1.9 billion pounds).

In Europe, the largest single producer is Germany, followed by France. In France it is estimated that some 60% of the commercial plantings are 'Golden Delicious'. In the United Kingdom, sales of the leading commercially grown cultivars are 'Gala' and 'Royal Gala' (21%), 'Braeburn' (17%), 'Cox's Orange Pippin' (17%), 'Golden Delicious' (14%), 'Granny Smith' (11%), and other cultivars (15%). A fact should be noted, with some humility for British apple lovers. Although there are more than 2,000 distinct cultivars of what we could call the premier division of dessert apples in Britain (Janes 1998), only 'Cox's Orange Pippin' (whose pollen parent is unknown) is a native English raising, and the parentages of all the rest ('Gala' excepted, 'Kidd's Orange Red' × 'Golden Delicious'), including 'Bramley's Seedling', are either unknown or contentious. And elsewhere, other major unknowns include 'Jonathan' (perhaps a seedling from 'Esopus Spitzenburg' in late 18th century New York State; Hanson 2005), 'McIntosh', and 'Winesap' from the United States, and 'Belle de Boskoop' (also called 'Schöne von Boskoop', depending on which side of the border one is on, both more than likely coming from the same orchard) and 'Ingrid Marie' from continental Europe. The finest efforts of the world's plant breeders have achieved some success, but breeding apples for worldwide commercial penetration is still a high-risk enterprise.

Apple wood

The virtues of the apple extended even further—to its timber. Although not often found in large log sizes, apple wood was used by the early settlers of North America for mauls, gavels, mallets, wedges, handsaw and other tool handles, machinery bearings, the teeth of mill- or waterwheels, shoemakers' lasts, and chair rockers. It found rare use in flutes and other musical instruments, bowls, toys, and cabinets. It was, in the last resort, one of the most favored fire logs, imparting a delightful fragrance to a warming room (Plate 26). This multitude of applications, but only from usable-sized timber of mature trees, ensured that apple seedlings were not ruthlessly grubbed out as worthless weed trees wherever they appeared. This range of long-term virtues would have ensured that very large numbers of randomly pollinated seedlings survived, often to extreme old age whatever their immediate culinary value, building a huge, long-lived gene pool of potential cultivars, as in the apple's homeland of Inner and Central Asia.

Apple hybrids

Thomas Andrew Knight knew a good deal about cider and was one of the first truly scientific pioneers of plant breeding (Knight 1797). Long before Gregor Mendel, he came extremely close to discovering the particulate nature of inheritance. He was unlucky enough to have selected, time and again, what we would now call polygenic characteristics controlled not by a single gene, like Mendel's classical round or wrinkled pea characters, but instead by multigenic, linked systems that are difficult to analyze.

Knight was well aware of the ravages of fruit diseases such as apple scab fungus (caused by the fungus *Venturia inaequalis*) and codling moth (*Cydia pomonella*, often referred either to the genus *Enarmonia* or *Carpocapsa*). He attempted to hybridize cultivars of *Malus pumila*, for example, 'Doctor Harvey', which is highly susceptible to scab, with resistant forms of the recently introduced Siberian crab (*M. baccata*, Plate 5). Several hybrid offspring were initially successful, notably 'Siberian Bitter Sweet' and 'Siberian Harvey'. 'Siberian Harvey' was depicted by Hooker (1818; and 1989), and both cultivars were listed with commendation by Hogg (1884), but neither seems to have survived.

Conventional apple breeding long concentrated on a very small number of cultivars (Hampson and Kemp 2003), for example:

'Allington Pippin' = 'King of the Pippins' (Plate 6) × 'Cox's Orange Pippin'
'Charles Ross' = 'Peasgood Nonsuch' × 'Cox's Orange Pippin'
'Cox's Orange Pippin', from a seed of 'Ribston Pippin' (Plate 8)
'Discovery' (Plates 6 and 7) = 'Worcester Pearmain' × 'Beauty of Bath'
'Falstaff' = 'James Grieve' × 'Golden Delicious'
'Fiesta' = 'Cox's Orange Pippin' × 'Ida Red'
'Fuji' = 'Ralls Janet' × 'Delicious'
'Gala' = 'Kidd's Orange Red' × 'Golden Delicious'
'Greensleeves' = 'James Grieve' × 'Golden Delicious'
'Jupiter' = 'Cox's Orange Pippin' × 'Starking'
'Katy' or 'Katja' = 'James Grieve' × 'Worcester Pearmain'
'Kidd's Orange Red' = 'Cox's Orange Pippin' × 'Delicious'
'King George V', from a seed of 'Cox's Orange Pippin'
'Laxton's Superb' = 'Wyken Pippin' × 'Cox's Orange Pippin'
'Laxton's Triumph' = 'King of the Pippins' (Plate 6) × 'Cox's Orange Pippin'
'Malling Kent' = 'Cox's Orange Pippin' × 'Jonathan'
'Merton Pippin' = 'Cox's Orange Pippin' × 'Sturmer Pippin' (Plate 11)
'Merton Worcester' = 'Cox's Orange Pippin' × 'Worcester Pearmain'
'Michaelmas Red' = 'McIntosh' × 'Worcester Pearmain'
'Newport Cross' = 'Devonshire Quarrenden' × 'Cox's Orange Pippin'
'Pink Lady' = 'Golden Delicious' × 'Lady Williams'
'Sunset', from a seed of 'Cox's Orange Pippin'
'Tydeman's Early Worcester' = 'McIntosh' × 'Worcester Pearmain'
'Tydeman's Late Orange' = 'Laxton's Superb' × 'Cox's Orange Pippin'
'William Crump' = 'Cox's Orange Pippin' × 'Worcester Pearmain'
'Winston' (Plate 9) = 'Cox's Orange Pippin' × 'Worcester Pearmain'

(The word *pearmain,* in 'Worcester Pearmain', is believed to be an allusion to the pear-shaped fruit.) Many presumed parentages are shown to be incorrect when modern DNA methods are used for analysis (Kitahara et al. 2005). For example, 'Honeycrisp', the important cold-hardy cultivar raised in Minnesota (called 'Honeycrunch' in Europe), was thought to be a cross between 'Honeygold' and 'Macoun', but neither is in fact a parent, though 'Keepsake' seems to be one (Cabe et al. 2005). The use of a limited number of cultivars for breeding combined with the fact that large-scale apple plantings comprise fewer than a dozen cultivars create an incestuous situation that is probably unsustainable in the long term (Noiton and Alspach 1996).

Ornamental crabs

In China, Japan, and to a certain extent Europe but markedly so in the United States, plant breeders have concentrated on a number of the wild species of *Malus* and derived from them large shrubs or small trees of great and lasting ornamental value. Many such ornamental cultivars are referred to as crabs. The word *crab* comes from the Old English *crabbe,* meaning bitter or sharp tasting. But does "crab apple," though widely used, have a precise meaning?

In western Europe, "crab apple" is often used to refer to *Malus sylvestris.* Probably one of the first species of *Malus* to be recognized as a "flowering crab" was *M. spectabilis.* This, called the Asiatic apple, *hai-t'ang,* or Chinese flowering crab, with its dark rose-red buds opening to beautiful, bright pink flowers, was already being cultivated in the time of the T'ang dynasty, A.D. 618–907. It was much grown in the grounds of the imperial palace in Beijing, and the small but acid-sharp fruits were eaten in a candied form.

A small number of hybrid cultivars were favored sources of conserves called crab apple jelly—'John Downie' in Britain, 'Chestnut Crab', raised in Minnesota before 1921, and 'Dolgo', raised in St. Petersburg, Russia, before 1917—but because of the labor-intensive nature of its production, that is now, sadly, a rare sight at the tea table.

What is the parentage of 'John Downie' and other, more widely planted ornamental crabs such as 'Cheal's Scarlet', 'Dartmouth', and 'The Siberian'? The literature seems to be silent on this matter. The well-known ornamental, Japanese flowering crab (*Malus ×floribunda*), may have originated in Japan before 1862, but its parentage, too, is unknown (Fiala 1994), though it could be *M. sieboldii* (now included in *M. toringo*) × *M. baccata.* With its profuse, distinctly fragrant, abun-

dant flowers, its remarkable resistance to disease, and its capacity to produce so
many outstanding hybrids, Japanese flowering crab must be among the most
widely planted ornamental crabs. As is true of most such crabs, *M.* ×*floribunda* has
small, long-pediceled, bird-distributed, yellow-brown fruits of no culinary value.

'Niedzwetzkyana'

With the exception of 'Niedzwetzkyana' (Plate 12), large-fruited *Malus pumila* is
not an important contributor to ornamental crab breeding. Due to its striking
appearance, this particularly dark form of *M. pumila* was introduced into the
British Isles in 1894 and then, from another collection, into the United States in
1897. Niels Hansen was a plant hunter and a grower at the South Dakota Agricultural Experiment Station in Brookings who obtained his seminal importation
from a Mr. Niedzwetzky of Almaty, Kazakhstan, and crossed that plant with *M.
baccata*, giving rise to the Rosybloom group of ornamental crabs (Fiala 1994).
Many Rosybloom cultivars have a striking crimson skin and flesh as well as their
most outstanding feature: brilliant crimson petals. In their original wild forms
and as modern hybrids, they are very much a part of the U.S. ornamental crab
program. The flesh of many, if not most, Rosybloom cultivars is attractively
spotted, tinged, suffused, or even wholly crimson. Nevertheless, they are mostly
inedible. It is also possible that the ornamentals 'Wisley Crab' and 'Harry Baker'
may have originated from the English introduction, but this has been neither
confirmed nor refuted by microsatellite or similar DNA analyses. It was claimed
by I. V. Michurin that the ornamental 'Belfleur Krasnyi', raised in Murmansk,
Russia, in 1914, also had *M. pumila* 'Niedzwetzkyana' as one of its parents (Fayers
2002). Again, this speculation needs testing by modern methods, and statements
as to parentage in the pre–DNA analysis literature should be treated with caution.
    'Niedzwetzkyana' seems to be a rare color mutant of *Malus pumila* found at
very low frequency throughout the Tian Shan and seen as fruits for sale in the
Asian markets. Although sometimes treated as a distinct species, it is no more
than a variant of *M. pumila*. The leaves, petioles, stems, and even the young wood
are purplish red, but the rose-purple flowers are particularly spectacular. The
fruit is of medium size, conical and ribbed, with a dark, claret red flesh, often
crimson throughout the whole pulp. The flesh is juicy but virtually tasteless and
of no culinary importance. The plant is very rarely found now in the wild because
of its ornamental value. Even at the seedling stage its future promise can be detected, and local villagers or herdsmen dig up young plants and sell them at a
premium in markets. Probably both deliberately and accidentally, *M. pumila*
'Niedzwetzkyana', with its striking flowers and fruits and capacity to hybridize,

has entered into ornamental breeding. It seems very likely that some of the ornamental crabs of our gardens derive, in part, and entirely accidentally, from it.

There is an early but not wholly convincing explanation for the red pigment of the apple flesh typical of 'Niedzwetzkyana'. It centers on Micah Rood, a prosperous farmer at Franklin, Pennsylvania. One day in 1693 a peddler of jewelry called at the farmstead and next day was found murdered under an apple tree in Rood's orchard. The farmer was never prosecuted, but the following autumn all the apples of the fatal tree were stained red in their flesh. Shortly afterward, in unexplained circumstances, the farmer was found dead. The phenomenon is now called Micah Rood's Curse (Room 1998).

A more likely explanation for the red-stained flesh in some apples is as follows. As far as can be ascertained, among the diverse *Malus pumila* trees in the fruit forest there are individuals with a multiallelic form of control for rich anthocyanin production, hence the bright color of a phase 1 apple. When these multiple loci come together in a particular combination and high frequency, the result is not only a brilliant red fruit skin, with usually red foliage, young stems, stem tissue, bark, and buds, but also occasionally partly or completely red fruit flesh. This is characteristic of the cultivars 'Beauty of Bath', 'Bloody Ploughman', 'Discovery' (Plates 6 and 7), 'McIntosh Red', 'Reinette Rouge Étoilée', 'Spartan', and 'Ten Commandments'. 'Niedzwetzkyana' is likely, therefore, simply manifesting an extreme expression of these particular gene sequences.

CHAPTER 8

A dénouement

A WEALTHY PERSIAN LANDOWNER at the time of Cyrus the Great, about 2,500 years ago, built and planted his paradise garden (see page 109). Therein, with his family and guests, he enjoyed apples that did not differ in any significant way, either in size, appearance, taste, the manner of their production, or their genetic identity, from those that we pick from the trees in our own gardens or select from supermarket shelves.

About 120 million years ago, what we now call flowering plants began to be preserved in the fossil record. They slowly began to replace, but never completely to overwhelm, other plant forms such as mosses, ferns, horsetails, cycads, and conifers. Somewhere, perhaps in the region that is now central to southern China—but almost certainly it will never be known exactly where on an ever-changing, restless Earth—and perhaps about 50 million years ago, plants appeared that if they lived today would be recognized as belonging to the rose family.

Perhaps 10 million years ago, an early form of the apple became trapped on a series of rising mountain ranges on the emerging landmass of Inner and Central Asia. Trees of this eo-apple carried cherry-sized, long-stemmed fruits, probably distributed by birds. Most wild apples, including those native to North America, are sour or at least astringent to the taste.

About 1.75 million years ago, glaciers began to descend into western and central Europe. But the Tian Shan was never glaciated, and what is more, the vegetation of the mountain slopes and foothills was continuously invigorated both by water from the high snowpack and its ever-rising, crumbling and breaking, diverse rock formations brought about by earthquakes. About the same time as the glaciers began their descent, deserts began to form around the Tian Shan, particularly its eastern arc. These events closed the Tian Shan off from the rest of Asia and from Europe. But in the mountain refuge, interactions between the plant pioneers, including the eo-apple, and the animals, particularly bears, slowly brought about remarkable changes in this eo-apple, resulting in the plants we now identify as *Malus pumila*. The geographical isolation of the Tian Shan and intense selection pressure also seem to have built an immense variability into the population of *M. pumila*—every apple plant in the Tian Shan is different in some small way (Plate 21), but many bear sweet apples. Also inherent in this species is the capacity, wherever or whenever even small populations of the trees are moved to other lands, to regenerate that astonishing variability.

About 10,000 years ago, or perhaps a little longer, humans, who had been drawn away from their immediate ancestors for at least 2 million years, were subjected to another dramatic change. Some hunter-fisher-gatherer peoples, originally with no fixed abode, made the transition, the Neolithic Revolution, to a settled existence based on the planting of crops and the storage of food. By coincidence, the first two sites of this transition were in the valley of the Tigris and Euphrates, in parts of modern-day Iraq, Syria, and Turkey, and to the east in the valley of the Chang, or Yangtze, of modern-day China (Map 9). The two emerging civilizations were roughly 6,000 km (3,700 miles) apart but lay approximately equidistant, west and east, respectively, from the Tian Shan. Trading routes slowly began to form between these Neolithic communities, based almost certainly on ancient, grazing-animal migration routes. As the summer's heat became fierce

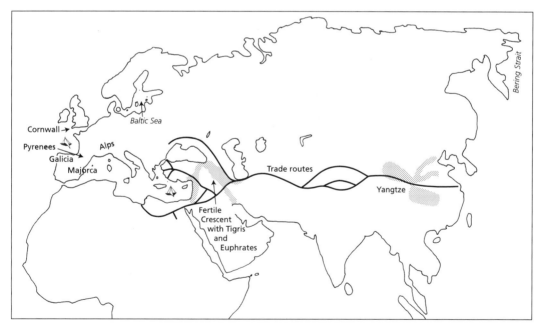

MAP 9. Agriculture, the Neolithic Revolution, started in the Middle East in the so-called Fertile Crescent with the rivers Tigris and Euphrates, followed shortly by similar developments in the valley of the Chang, or Yangtze, to the east.

and the grass and herbs withered, the shorter, lower, desert routes would have been abandoned. Instead, the longer, higher, but cooler mountain routes were used. These led through the pristine fruit forest of apples, pears, apricots, pistachios, plums, and walnuts that had nutritious fruits attractive to human beings as well as bears and other animals, mostly on the cooler, north-facing slopes of the ranges.

About 7,000 years ago, the horse, a native of the steppes north of the Tian Shan, was domesticated. Such a development speeded the movement of the expanding human communities of the region. It also accelerated wider distribution of apples because, unlike in a camel, whose chewing and digestion destroy seeds, an apple pip passes undamaged through a horse's gut and is deposited in a fertile growth medium. The sweet apple of the Tian Shan began to move west.

In Babylon, about 3,800 years ago, the grafting of fruit trees was perfected. This revolutionary technique—instant domestication, it has been called—allowed the choicest specimens of the forest to be transported to the newly emerging gardens, first of Persia and then of Greece and Rome. War and conquest hastened their spead. The Romans were possibly the first to bring the whole apparatus of grafting—stocks, scions, and a range of cultivars—to Britain. But evidence is accumulating that at least seedling trees of the sweet apple may already have been present in western Europe. From its widened Eurasian base, the apple was carried across the seas. The rest is history.

To this day, the diverse colors, shapes, sizes, tastes, and textures of the many thousands of cultivars of the apple-growing countries of the world can still be matched by the diversity in the Tian Shan. In harsh contrast, the kiwi fruit (*Actinidia deliciosa*) industry, for example, is based almost entirely on just one cultivar, 'Hayward'.

Figs, both Smyrna and Sycamore types, grapes, pomegranates (Zohary and Spiegel-Roy 1975), mulberries, gooseberries, and some Rosaceae such as raspberries and strawberries from the cool temperate zone can be grown from cuttings, but for most of the tribe Pyreae and their allies, including apple, pear, quince, plum, and cherry, virtually none will strike. And unlike, for instance, the banana, which is virtually sterile across the world and which incidentally cannot easily be grafted, the apple, whether grafted or on its own roots, has retained a vital safety net with its potential ability to grow from seed.

As has frequently been proposed, hybridization between the sweet apple and the European species of *Malus* could have taken place. But DNA analyses, conversations with apple breeders, and the work of Coart et al. (2003) indicate that this must be very rare. Perhaps the nearly complete absence of endosperm in the seed of *M. pumila* hinders such hybridization. Grafting has enabled outstanding individual cultivars trees to be preserved through cloning, with little if any change over long periods, in some cases perhaps as much as 2,000 years. Nevertheless, every hedgerow and thicket in Europe and many roadsides in the Americas testify to the extraordinary sexual reproduction potential of *M. pumila*. Except for the most minute DNA base-pair duplication, roughly the equivalent of one typographical error in this whole book, the original wild apple seems to have changed in almost no degree since the dawn of agriculture and over perhaps 10,000 years.

The future of the apple

The production of practically all the economically important grains and legumes, many of the fruits, and a vast swath of other crops and ornamentals has passed out of the control of the individual grower. It becomes ever more difficult for the individual farmer or gardener, should he or she so wish, to save seeds for a future generation. A combination of the Green Revolution, agroindustrial control exercised by genetic modification, and bureaucratic diktat ensures that the genetic basis of our food becomes more restricted by the day.

But the apple remains a determined, effective, subversive influence. Through the whole growing season, the supermarket shelves of the world display for good commercial reasons probably fewer than 50 apple cultivars. Yet in the orchards of abandoned homesteads in upper New York State, along the ancient trackways of England, and in thickets and on cliff slopes and other marginal land in the whole of the temperate world, seedling apples of unknown parentage constantly arise. Most of these are of little value except for pig food or rough cider, but here and there a rare, elite individual will emerge. Inexplicably, these chance occurrences seem to take place very rarely in similarly cultivated pears or plums.

Via simple techniques developed by Bronze Age farmers in the valley of the Tigris and Euphrates 3,800 years ago, and easily learned, these new genotypes can be spread to the four corners of the earth. Their seedlings lie beyond the control of large-scale commerce, governmental pressure, or the boardrooms of agribusiness.

Classification and distribution of apple species

Here are listed the wild species of the genus *Malus*, adapted from Phipps et al. (1990), and their geographical ranges. Keep in mind that such a listing is fraught with problems—the genus is in need of a modern taxonomic monograph, and accounts in contemporary floristic treatments (*Flora Europaea, Flora of Turkey, Flora U.S.S.R., Flora of China*, and *Flora of Japan*) are at odds with one another.

section *Malus*
    series *Malus*

M. asiatica Nakai	Xinjiang Uygur and Liaoning to Yunnan
M. chitralensis Vassilczenko	western Pakistan
M. dasyphylla Borkhausen	central Europe, Balkans
M. kirghisorum Alexander & Andrej Theodorov	Plate 4; Iran, Central Asia
M. montana Uglitskikh	Central Asia
M. praecox (Pallas) Borkhausen	Russia
M. prunifolia (Willdenow) Borkhausen	plumleaf crab; northeastern China
M. pumila Miller, including *M. domestica* (Borkhausen) Borkhausen, *M. niedzwetzkyana* Dieck ex Koehne, and *M. sieversii* (Ledebour) M. Roemer	sweet apple, Paradise apple (Plates 1, 2, and 12); western Xinjiang Uygur and now worldwide
M. spectabilis (Aiton) Borkhausen	Asiatic apple, Chinese flowering crab; Yunnan, eastern China
M. sylvestris (Linnaeus) Miller, including *M. orientalis* Uglitskikh	European crab (Plate 3); Europe, Caucasus Mountains, Iran

M. turkmenorum Juzepczuk   Central Asia
   & Popov
series *Baccatae*
   M. baccata (Linnaeus) Borkhausen,   Siberian crab (Plate 5); northeastern
      including *M. mandshurica*   China, Russian Siberia, and as far
      (Maximowicz) Komarov   south as Kazakhstan in the
                                         Tian Shan

   M. halliana Koehne   central and southern China
   M. hupehensis (Pampanini)   central and eastern China
      Rehder
   M. pallasiana Juzepczuk   Mongolia, Far Eastern Russia
   M. rockii Rehder, conspecific with   Bhutan, Tibet, northwestern Yunnan
      M. sikkimensis of
      southwestern Sichuan?
   M. sachalinensis Juzepczuk   Sakhalin
   M. sikkimensis (Wenzig) Koehne   eastern Himalaya
   M. spontanea (Makino) Makino   Japan
section *Sorbomalus*
series *Sieboldianae*
   M. toringo (Siebold) de Vriese,   Japan, Korea, China
      including *M. sargentii* Rehder
      and *M. sieboldii* (Regel) Rehder
series *Florentinae*
   M. florentina (Zuccarini)   Italy, Balkans, Greece, Turkey
      Schneider
series *Kansuenses*
   M. fusca (Rafinesque) Schneider   Oregon crab; Alaska to California
   M. kansuensis (Batalin) Schneider   Central China
   M. komarovii (Sargent) Rehder   Jilin, China, North Korea
   M. toringoides (Rehder) Hughes   Gansu, southern Sichuan, eastern Tibet
   M. transitoria (Batalin) Schneider   north-central China
series *Yunnanenses*
   M. honanensis Rehder   north-central China
   M. ombrophila Handel-Mazzetti   northwestern Yunnan, southwestern
                                         Sichuan
   M. prattii (Hemsley) Schneider   northwestern Yunnan, southwestern
                                         Sichuan
   M. yunnanensis (Franchet)   south-central China, Myanmar
      Schneider

section *Chloromeles*

    M. angustifolia (Aiton) Michaux   southern crab apple; southern North
            America

    M. bracteata Rehder   southern United States

    M. coronaria (Linnaeus) Miller   sweet crab; eastern North America

    M. glabrata Rehder   Biltmore crab; eastern North America

    M. glaucescens Rehder   Dunbar crab; eastern North America

    M. ioensis (Wood) Britton   prairie crab; central North America

    M. lancifolia Rehder   eastern North America

    M. platycarpa Rehder   bigfruit crab; eastern North America

section *Docyniopsis*, sometimes treated
        as the genus *Docyniopsis*
        (Schneider) Koidzumi

    M. doumeri (Bois) A. Chevalier,   southern China, Laos, Vietnam
        including *M. formosana*
        (Hayata) Kawakami &
        Koidzumi, and *M. melliana*
        (Handel-Mazzetti) Rehder

    M. tschonoskii (Maximowicz)   Japan
        Schneider

section *Eriolobus*, sometimes treated
        as the genus *Eriolobus*
        (de Candolle) M. Roemer

    M. trilobata (Labillardière)   eastern Mediterranean
        Schneider

References

Akhmetov, A. K. 1998. *History of Kazakstan: Essays.* Ministry of Science–Academy of Sciences of the Republic of Kazakstan Institute of History and Ethnology. Gylym, Almaty.

Aldasoro, J. J., C. Aedo, and C. Navarro. 2005. Phylogenetic and phytogeographic relationships in Maloideae (Rosaceae) based on morphological and anatomical characters. *Blumea* 50: 3–52.

Allen, T. B. 1996. The Silk Road's lost world. *National Geographic* 189 (March): 44–51.

Amherst, A. M. T. 1894. A fifteenth-century treatise on gardening. By "Mayster Ion Gardener." *Archaeologia* 54: 157–172.

Amherst, A. 1895. *A History of Gardening in England.* Quaritch, London.

Anderson, E. 1952. *Plants, Man and Life.* Little Brown, Boston.

Ashmole, E. 1618. *Diary of John Tradescant, the Elder, Voyage to Siberia in 1618.* Bodleian Library manuscript 824, Oxford. (a full transcription is given by Leith-Ross 1998)

Austen, R. 1657. *A Treatise of Fruit-Trees Shewing the Manner of Grafting, Setting, Pruning, and Ordering of Them in All Respects: According to Divers New and Easy Rules of Experience; Gathered in the Space of Twenty Years.* Oxford. (1st edition 1653, a later edition 1665)

Austen, R. 1658. *Observations upon Some Part of Sr Francis Bacon's Naturall History As It Concernes, Fruit-Trees, Fruits, and Flowers.* Oxford.

Austen, R. 1676. *A Dialogue, or Familiar Discourse, and Conference Betweene the Husbandman, and Fruit-Trees.* Oxford.

Ball, W. 1998. Following the mythical road. *Geographical: The Royal Geographical Society Magazine* 70: 18–23.

Barber, E. W. 1999. *The Mummies of Ürümchi.* Macmillan, London.

Barnes, T. 1759. *A New Method of Propagating Fruit-Trees, and Flowering Shrubs.* London.

Bauhin, J. 1598. *Historia Plantarum Universalis, Nova, et Absolutissima, cum Consensu et Dissensu Circa Eas* [a universal and complete history of plants, with agreement and dissension therein]. D. Chabrée and F. L. von Graffenried, editors. Yverdon, Switzerland.

Bazeley, B. 1991. Pips that made history. *Country Life* (23 May): 90–91.

Beale, J. 1653. *A Treatise on Fruit Trees Shewing their Manner of Grafting, Pruning, and Ordering, of Cyder and Perry, of Vineyards in England.* Oxford.

Beale, J. 1664. Aphorisms concerning cider, pages 21–29 in J. Evelyn, *Sylva.* London.

Becerra Velásquez, V. L., and P. Gepts. 1994. RFLP diversity of common bean (*Phaseolus vulgaris*) in its centres of origin. *Genome* 37: 256–263.

Beeton, I. 1861. *The Book of Household Management.* S. O. Beeton, London. (2000 reprint edited by N. Humble in *Oxford World's Classics*, Oxford University Press)

Beresford, J., editor. 1924. *The Diary of a Country Parson* (Parson James Woodforde), 5 vols. Clarendon Press, Oxford.

Bock, H. 1546. *Kreüter Bůch.* Strasbourg. (other editions 1539, 1552, 1560)

Bökönyi, S. 1974. *The Przevalsky Horse,* translated from the Hungarian by Lili Halápy. Souvenir Press, London.

Boré, J. M., and J. Fleckinger. 1997. *Pommiers à Cidre: Variétés de France* [cider apples: the varieties of France]. INRA Éditions, Paris.

Borkhausen, M. B. 1803. *Theoretisch-praktisches Handbuch der Forstbotanik und Forsttechnologie* [theoretical and practical handbook of forest botany and forest technology], Vol. 2. Heyers, Giessen und Darmstadt.

Bošković, R., and K. R. Tobutt. 1999. Correlation of stylar ribonuclease isoenzymes with incompatibility alleles in apple. *Euphytica* 107: 29–43.

Boufford, D. E., and S. A. Spongberg. 1983. Eastern Asian–eastern North American phytogeographical relationships—A history from the time of Linnaeus to the twentieth century. *Annals of the Missouri Botanical Garden* 70: 423–439.

Bradley, R. 1724. *New Improvements of Planting and Gardening, both Philosophical and Practical,* 4th edition. London and Dublin. (other editions 1717–1726)

Brandenburg, W. A. 1991. The need for stabilized plant names in agriculture and horticulture, pages 23–31 in D. L. Hawksworth, editor, Improving the stability of names: Needs and options. *Regnum Vegetabile* 123.

Bretshneider, E. V. 1875. *Notes on Chinese Mediaeval Travellers to the West.* Shanghai.

Broecker, W. S., and T. Liu. 2001. Rock varnish: Recorder of desert wetness? *Geological Society of America Today* 11: 4–11.

Browićz, K. 1970. *Malus florentina*—Its history, systematic position and geographic distribution. *Fragmenta Floristica Geobotanica* 16: 1–83.

Brown, S. K., and K. E. Maloney. 2003. Genetic improvement of apple: Breeding, markers, mapping, and biotechnology, pages 31–60 in D. C. Ferree and I. J. Warrington, editors, *Apples: Botany, Production, and Uses.* CABI Publishing, Cambridge, Massachusetts.

Browning, F. 1999. *Apples: The Story of the Fruit of Temptation.* Allen Lane, London.

Bugnon, L., editor. 1995. Les croqueurs des pommes [the apple crunchers]. *Bulletin de Liaison* 67 (premier trimestre): 29. Association Nationale des Croqueurs de Pommes, Belfort.

Bulliet, R. W. 1975. *The Camel and the Wheel.* Harvard University Press, Cambridge.

Bultitude, J. 1983. *Apples: A Guide to the Identification of International Varieties.* Macmillan, London.

Bunyard, E. A. 1918. Cotton's "Planter's Manual." *Gardeners' Chronicle,* ser. 3, 63 (27 April): 174–175.

Bunyard, E. A. 1920. *A Handbook of Hardy Fruits More Commonly Grown in Great Britain: Apples and Pears.* Murray, London.

Bunyard, E. A. 1933. *The Anatomy of Dessert: With a Few Notes on Wine.* Chatto and Windus, London.

Bunyard, G., and O. Thomas. 1906. *The Fruit Garden,* 2nd ed. Country Life, London.

Buttenschon, R. M., and J. I. Buttenschon. 1998. Population dynamics of *Malus sylvestris* stands in grazed and ungrazed, semi-natural grasslands and fragmented woodlands in Mols Bjerge, Denmark. *Annales Botanici Fennici* 35: 233–246.

Cabe, P. R., A. Baumgarten, K. Onan, J. J. Luby, and D. S. Bedford 2005. Using microsatellite analysis to verify breeding records: A study of 'Honeycrisp' and other cold-hardy apple cultivars. *HortScience* 40: 15–17.

Cable, M., and F. French. 1936. *George Hunter: Apostle of Turkestan.* China Inland Mission, London.

Cable, M., and F. French. 1942. *The Gobi Desert.* Hodder and Stoughton, London. (1984 facsimile, Virago Press)

Calhoun, C. L., Jr. 1995. *Old Southern Apples.* McDonald and Woodward, Blacksburg, Virginia.

Campbell, C. S., M. J. Donoghue, and B. G. Baldwin. 1995. Phylogenetic relationships in Maloideae (Rosaceae): Evidence from sequences of the internal transcribed spacers of nuclear ribosomal DNA and its congruence with morphology. *American Journal of Botany* 82: 903–918.

Candolle, A. de. 1885. *Origin of Cultivated Plants.* Appleton, New York.

Cardia, P., S. H. Li, and N. Ferrand. 2002. The azure-winged magpie *Cyanopica cyanus* in Europe: A native or introduced species? *Phylogeography in Southern European Refugia: Evolutionary Perspectives on the Origins and Conservation of European Biodiversity.* CECA-ICETA, University of Porto, 11–15 March 2002, poster presentation, Vairão, Portugal.

Carswell, J. 2000a. An odyssey in blue and white. *Cornucopia* 5(25): 36–39.

Carswell, J. 2000b. *Blue and White: Chinese Porcelain Around the World.* British Museum Press, London.

Choy Leng Yeong. 2005. State apple farmers struggle as prices fall. *Seattle Post-Intelligencer* (20 August 2005): C1, C8.

Church, A. H. 1981. The botany of the garden in Eden, pages 237–245 in D. J. Mabberley, editor, *Thalassiophyta and Other Essays of A. H. Church.* Clarendon Press, Oxford.

Clark, M. 2003. *Apples: A Field Guide.* Whittet Books in association with Brogdale Horticultural Trust, Cotton, Ipswich.

Clutton-Brock, J. 1981. *Domesticated Animals from Early Times.* Heinemann and British Museum of Natural History, London.

Coart, E., X. Vekemans, M. J. M. Smulders, I. Wagner, J. van Huylenbroeck, E. van Bockstaele, and I. Roldán-Ruiz. 2003. Genetic variation in the endangered wild apple (*Malus sylvestris* (L.) Mill.) in Belgium as revealed by amplified fragment length polymorphism and microsatellite markers. *Molecular Ecology* 12: 845–857.

Collins, S. 1717. *Paradise Retriev'd: Plainly and Fully Demonstrating the Most Beautiful, Durable, and Beneficial Method of Managing and Improving Fruit-Trees.* London.

Cook, M. 1676. *The Manner of Raising, Ordering, and Improving Forrest-Trees: . . . .* London. (other editions to 1724)

Cooper, J. H. 2000. First fossil record of azure-winged magpie *Cyanopica cyanus* in Europe. *Ibis* 142: 150–151.

Cooper, J. H., and K. Voous. 1999. Birds, bones and biogeography: Iberian azure-winged magpies come in from the cold. *British Birds* 92: 659–665.

Copas, L. 2001. *A Somerset Pomona: The Cider Apples of Somerset.* Dovecote Press, Dorset.

Costa, L. M., J. F. Gutierrez-Marcos, and H. G. Dickinson. 2004. More than a yolk: The short life and complex times of the plant endosperm. *Trends in Plant Science* 9: 507–514.

Cotton, C. 1675. *The Planters Manual: Being Instructions for the Raising, Planting, and Cultivating All Sorts of Fruit-Trees.* London.

Cox, E. H. M. 1945. *Plant Hunting in China.* Collins, London.

Coxe, W. 1817. *A View of the Cultivation of Fruit Trees, and the Management of Orchards and Cider: With Accurate Descriptions of the Most Estimable Varieties of Native and Foreign Apples, Pears, Peaches, Plums and Cherries, . . .* Carey, Philadelphia. (1976 facsimile, Pomona Books, Rockton, Ontario)

Crossley, J. A. 1974. *Malus* Mill. apple, pages 531–534 in C. S. Schopmeyer, editor, *Seeds of Woody Plants in the United States.* Agricultural Handbook 450. U.S. Department of Agriculture Forest Service, Washington, D.C.

Cunliffe, B. 2001. *Facing the Ocean: The Atlantic and Its Peoples.* Oxford University Press.

Darlington, C. D. 1956. *Chromosome Botany.* Allen and Unwin, London.

Darlington, C. D., and A. A. Moffett. 1930. Primary and secondary chromosome balance in *Pyrus. Journal of Genetics* 22: 129–151.

Darlington, C. D., and A. P. Wylie. 1955. *Chromosome Atlas of Flowering Plants.* Allen and Unwin, London.

Darwin, C. 1845. *Journal of Researches into the Natural History and Geology of the Countries Visited During the Voyage of H.M.S. Beagle Round the World.* John Murray, London. (other, later editions)

Darwin, C. 1868. *The Variation of Animals and Plants Under Domestication,* 2 vols. John Murray, London.

Dennis, F. 2003. Flowering, pollination, and fruit set and development, pages 153–166 in D. C. Ferree and I. J. Warrington, editors, *Apples: Botany, Production, and Uses.* CABI Publishing, Cambridge, Massachusetts.

Dewey, J. F., R. M. Shackleton, C. Chengfa, and S. Yiyin. 1988. The tectonic evolution of the Tibetan Plateau. *Philosophical Transactions of the Royal Society of London,* Series A, 327: 379–413.

Diamond, J. M. 1991. The earliest horsemen. *Nature* 350: 275–276.

Dickson, E. E., S. Kresovich, and N. F. Weeden. 1991. Isozymes in North American *Malus* (Rosaceae): Hybridization and species differentiation. *Systematic Botany* 16: 363–375.

Domitzer, J. 1531. *Ein Neues Pflantzbüchlin, von Mancherley Artiger Propffung, und Beltzung der Bäum* [a new little plant book concerning various ways of propagating and growing trees]. Augsburg, Germany.

Donoghue, M. J., C. D. Bell, and J. Li. 2001. Phylogenetic patterns in northern hemisphere plant geography. *International Journal of Plant Science* 162 (6 suppl.): 541–552

Drope, F. 1672. *A Short and Sure Guid in the Practice of Raising and Ordering of Fruit-Trees.* Oxford.

Drower, M. S. 1969. The domestication of the horse, pages 471–478 in P. J. Ucko and G. W. Dimbleby, editors, *The Domestication and Exploitation of Plants and Animals.* Duckworth, London.

Dudek, M. G., L. Kaplan, and M. Mansfield King. 1998. Botanical remains from a seventeenth-century privy at the Cross Street Back Lot site. *Historical Archaeology* 32: 63–71.

Duhamel du Monceau, H. L. 1807–1835. *Traité des Arbres Fruitiers. Nouvelle Édition, Augmentée d'un Grand Nombre des Espèces des Fruits Obtenus de Progrès de la Culture* [a treatise of fruit trees, improved by the addition of a large number of species of fruits resulting from the advance of fruit culture], illustré par A. Poiteau et P. J. F. Turpin, 6 vols. Levrault, Paris et Strasbourg.

Dunemann, F., R. Kahnau, and H. Schmidt. 1994. Genetic relationships in *Malus* evaluated by RAPD 'fingerprinting' of cultivars and wild species. *Plant Breeding* 113: 150–159.

Dzhangaliev, A. D. 2003. The wild apple tree of Kazakhstan. *Horticultural Reviews* 29: 65–303.

Dzhangaliev, A. D., T. N. Salova, and P. M. Turekhanova. 2003. The wild fruit and nut plants of Kazakhstan. *Horticultural Reviews* 29: 305–371.

Ekwall, E. 1991. *The Concise Oxford Dictionary of English Place Names,* 4th edition. Clarendon Press, Oxford.

Eriksson, T., J. E. E. Smedmark, M. S. Kerr, L. A. Alice, R. C. Evans, and C. S. Campbell. 2005. Major clades in the evolution of Rosideae. *XVII International Botanical Congress—Abstracts:* 100. Vienna.

Evans, R. C., and C. S. Campbell. 2002. The origin of the apple subfamily (Maloideae; Rosaceae) is clarified by DNA sequence data from duplicated GBSSI genes. *American Journal of Botany* 89: 1,478–1,484.

Evelyn, J. 1664. *Sylva, or, a Discourse of Forest-Trees, . . . to Which Is Annexed Pomona; or, an Appendix Concerning Fruit-Trees in Relation to Cider . . . .* London. (other editions to 1825)

Evelyn, J. 1699. *Acetaria. A Discourse of Sallets.* London.

Evenari, M. 1949. Germination inhibitors. *Botanical Reviews* 15: 153–194.

Farrer, R. 1926. *On the Eaves of the World,* 2 vols. Arnold, London.

Fayers, G. 2002. More news on the apples from Kazakhstan. *Flora Facts and Fables* 31: 10–12.

Ferree, D. C., and R. F. Carlson. 1987. Apple rootstocks, in R. C. Rom and R. F. Carlson, editors, *Rootstocks for Fruit Crops.* Wiley, New York.

Ferree, D. C., and I. J. Warrington, editors. 2003. *Apples: Botany, Production, and Uses.* CABI Publishing, Cambridge, Massachusetts.

Fiala, J. L. 1994. *Flowering Crabapples: The Genus Malus.* Timber Press, Portland, Oregon.

Fitzherbert, J. 1548. *The Boke of Husbandry,* 2nd edition. London. (1st edition 1523)

Fleming, P. 1934. *Travels in Tartary—One's Company and News from Tartary.* Reprint Society, London.

Fois, B., editor. 1981. *Il Capitulare de Villis* [from Charlemagne's capitulary edict, that is, the chapter concerning farms], manuscript 287. A. Giuffrè, Milan. (the original of the *Capitulare* is in the Herzog August Bibliothek at Wolfenbüttel under the shelf mark Cod. Guelf. 254 Helmst; a facsimile was published in 1971 by Carlrichard Brühl, Stuttgart)

Fok, K. W., C. M. Wade, and D. Parkin. 2002. Inferring the phylogeny of disjunct populations on the azure-winged magpie, *Cyanopica cyanus,* from a mitochondrial DNA sequence. *Phylogeography in Southern European Refugia: Evolutionary Perspectives on the Origins and Conservation of European Biodiversity.* CECA-ICETA, University of Porto, 11–15 March 2002, poster presentation, Vairão, Portugal.

Forsline, P. L. 1995. Adding diversity to the national apple germplasm collection: Collecting wild apples in Kazakhstan. *New York Fruit Quarterly* 3: 3–6.

Forsline, P. L., E. E. Dickson, and A. D. Dzhangaliev. 1994. Collection of wild *Malus, Vitis* and other fruit species genetic resources in Kazakhstan and neighboring republics. *HortScience* 29: 433. (abstract)

Forsline, P. L., H. S. Aldwinckle, E. E. Dickson, J. L. Luby, and S. C. Hokanson. 2003. Collection, maintenance, characterization, and utilization of wild apples of central Asia. *Horticultural Reviews* 29: 2–61.

Forster, E. S., and E. H. Heffner, editors and translators. 1979. *Lucius Junius Moderatus Columella on Agriculture,* Vols. 1–2, and *on Agriculture and Trees,* Vol. 3. With a recension of the text and an English translation by H. B. Ash. Loeb Classical Library, Harvard University Press, Cambridge, Massachusetts, and Heinemann, London.

Forsyth, W. 1802. *A Treatise on the Culture and Management of Fruit Trees; . . .* Longmans and Rees, London. (other editions to 1824)

Fortune, R. 1847. *Three Years' Wanderings in the Northern Provinces of China, Including a Visit to the Silk, and Cotton Countries: With an Account of the Agriculture and Horticulture of the Chinese,* 2nd edition. Murray, London.

Frary, A., T. C. Nesbitt, A. Frary, S. Grandillo, E. van der Knaap, B. Cong, J. Liu, J. Meller, R. Elber, K. B. Alpert, and S. D. Tansley. 2000. A quantitative trait locus key to the evolution of tomato fruit size. *Science* 289: 85–88.

French, R. K. 1982. *The History and Virtues of Cyder.* St. Martin's Press, New York, and Hale, London.

Gamkrelidze, T. V., and V. V. Ivanov. 1984. *The Indoeuropean Language and the Indoeuropeans.* Mouton de Gruyter, Berlin.

Geibel, M., K. J. Dehmer, and P. L. Forsline. 2000. Biological diversity in *Malus sieversii* populations from Central Asia. *Acta Horticulturae* 538: 43–49.

Gerard, J. 1597. *The Herball or Generall Historie of Plantes.* London.

Gerard, J. 1633. *The Herball or Generall Historie of Plantes . . . very much enlarged and amended by Thomas Johnson.* London. (other editions to 1636)

Gopher, A., S. Abbo, and S. Lev-Yadun. 2001. The "when," the "where" and the "why" of the Neolithic revolution in the Levant, pages 49–62 in M. Budja, editor, *Documenta Praehistorica* 28, *Neolithic Studies* 8. Ljubljana.

Grant, V. 1971. *Plant Speciation*. Columbia University Press, New York.

Grassi, F., G. Morico, and A. Sartori. 1998. *Malus* and *Pyrus* germplasm in Italy, pages 45–49 in L. Maggioni, R. Janes, A. Hayes, T. Swinburne, and E. Lipman, editors, *Report of a Working Group on Malus / Pyrus*. Meeting 15–17 May 1997, Dublin. International Plant Genetic Resources Institute.

Gray, T. 1995. *Devon Household Accounts, 1627–1659*. Devon and Cornwall Record Society. BPC Wheatons, Exeter.

Greenberg, J. H., and M. Ruhlen. 1992. Linguistic origins of native Americans. *Scientific American* November: 60–65.

Grew, N. 1675. *The Comparative Anatomy of Trunks, Together with an Account of Their Vegetation Grounded Thereupon*, London.

Grindon, L. H. 1885. *Fruits and Fruit Trees. Home and Foreign. An Index to the Kinds Valued in Britain*. Palmer and Howe, Manchester.

Groen, J. van der. 1669, 1681. *Le Jardinier du Pays-Bas* [the gardener of the Low-Countries]. Vleugart, Brussels.

Guilford, P., S. Prakash, J. M. Zhu, E. Rikkerink, S. Gardiner, H. Bassett, and R. Forster. 1997. Microsatellites in *Malus ×domestica* (apple): Abundance, polymorphism and cultivar identification. *Theoretical and Applied Genetics* 94: 249–254.

Haines, R. 1684. *Aphorisms upon the New Way of Improving Cyder, or Making Cyder-Royal, Raising and Planting of Apple-Trees*. London.

Hall, A. D., and M. B. Crane. 1933. *The Apple*. Hopkins, London.

Hampson, C. R., and H. Kemp. 2003. Characteristics of important commercial apple cultivars, pages 61–89 in D. C. Ferree and I. J. Warrington, editors, *Apples: Botany, Production, and Uses*. CABI Publishing, Cambridge, Massachusetts.

Hanski, I., and Y. Cambefort, editors. 1991. *Dung Beetle Ecology*. Princeton University Press.

Hanson, B., editor. 2005. *The Best Apples to Buy and Grow*. Brooklyn Botanic Garden, New York.

Harada, T., K. Matsukara, T. Sato, R. Ishikawa, M. Niizeka, and K. Saito. 1993. DNA RAPDs detect genetic variation and paternity in *Malus. Euphytica* 65: 87–91.

Harlan, J. R. 1992. *Crops and Man*. American Society of Agronomy, Madison, Wisconsin.

Harman, O. S. 2004. *The Man Who Invented the Chromosome*. Harvard University Press, Cambridge, Massachusetts.

Harris, S. A. 1999. RAPDs in systematics—A useful methodology? Pages 211–228 in P. M. Hollingsworth, R. M. Bateman, and R. J. Gornall, editors, *Advances in Molecular Systematics and Plant Evolution*. Taylor and Francis, London.

Harris, S. A., J. P. Robinson, and B. E. Juniper. 2002. Genetic clues to the origin of the apple. *Trends in Genetics* 18: 426–430.

Hartlib, S. 1645. *Discourse of Husbandrie used in Brabant and Flanders*. London.

Harvey, J. H. 1981. *Mediaeval Gardens*. Batsford, London.

Harvey, J. H. 1985. The first English garden book. *Garden History* 13(2): 83–101.

Harvey, J. H. 1992. Ibn Bassāl's *The Book of Agriculture*. Draft manuscript translation by the late John Harvey from the Spanish version by José Millás Vallicrosa based on the fragmentary Arabic text and the partial remains of the Castilian translation of the late 13th century preserved in the Biblioteca National de España (manuscript 10106). The whole of the surviving parts of the Castilian translation were printed by José María Millás Vallicrosa in *Al-Andalus* 13: 347–430 (1948) before the discovery of an Arabic text. The whole book, insofar as it survives, was published as *Libro de Agricultura,* edited, translated, and annotated by Villacrosa and Muhammad Azīmān (Tetuán, Morocco, Instituto Muley el-Hasan, 1955).

Hatton, R. G. 1917. Paradise apple stocks. *Journal of the Royal Horticultural Society* 42: 361–399.

Hearman, J. 1936. The Northern Spy as a rootstock when compared with other standardized European rootstocks. *Journal of the Pomological and Horticultural Sciences* 14: 246–275.

Hedrick, U., editor. 1919. *Sturtevant's Notes on Edible Plants*. Lyon, Albany, New York. (1972 reprint as *Sturtevant's Edible Plants of the World*, Dover, New York)

Hehn, V. 1902. *Kulturpflanzen und Haustiere in Ihrem Übergang aus Asien nach Griechenland und Italien Sowie in das Übrige Europa,* [cultivated plants and domesticated animals in their migration from Asia to Greece and Italy as well as to the rest of Europe], 7th edition, pages 613–616. Borntraeger, Berlin.

Henrey, B. 1975. *British Botanical and Horticultural Literature before 1800,* 3 vols. Oxford University Press.

Herrera, C. M. 1989. Frugivory and seed dispersal by carnivorous mammals, and associated fruit characteristics, in undisturbed Mediterranean habitats. *Oikos* 55: 250–262.

Hey, J. 2005. On the number of New World founders: A population genetic portrait of the peopling of the Americas. *Public Library of Science, Biology* 3: 965–975.

Hill, T. 1563. *A Most Briefe and Pleasaunte Treatyse, Teachynge Howe to Dress, Sowe, and Set a Garden*. London. (other editions to 1608)

Hitt, T. 1755. *A Treatise of Fruit-Trees*. London. (and other editions)

Hogg, R. 1851. *British Pomology or a History, Description, Classification and Synonymes of the Fruits and Fruit Trees of Great Britain: The Apple.* Groombridge, London.

Hogg, R. 1884. *The Fruit Manual: A Guide to the Fruits and Fruit Trees of Great Britain,* 5th edition. Journal of Horticulture Office, London.

Hogg, R., and H. G. Bull, editors. 1876–1885. *The Herefordshire Pomona, Containing Original Figures and Descriptions of the Most Esteemed Kinds of Apples and Pears.* Jakeman and Carver, Hereford.

Hokanson, S. C., J. R. McFerson, P. L. Forsline, W. F. Lamboy, J. J. Luby, A. D. Dzhangaliev, and H. S. Aldwinckle. 1997. Collecting and managing wild *Malus* germplasm in its center of diversity. *Horticultural Science* 32: 173–176.

Hokanson, S. C., W. F. Lamboy, A. K Szewc-McFadden, and J. R. McFerson. 1998. Microsatellite (SSR) markers reveal genetic identities, genetic diversity and relationships in a *Malus* ×*domestica* Borkh. core subset collection. *Theoretical and Applied Genetics* 97: 671–683.

Hokanson, S. C., P. L. Forsline, J. R. McFerson, W. F. Lamboy, H. S. Aldwinckle, J. L. Luby, and A. D. Dzhangaliev. 1999. Ex situ and in situ conservation strategies for wild *Malus* germplasm in Kazakhstan. *Acta Horticulturae* 484: 85–91.

Hokanson, S. C., W. F. Lamboy, A. K. Szewc-McFadden, and J. R. McFerson. 2001. Microsatellite (SSR) variation in a collection of *Malus* (apple) species and hybrids. *Euphytica* 118: 281–294, 363–372, 375.

Hooker, W. 1818. *Pomona Londinensis Containing Coloured Engravings of the Most Esteemed Fruits Cultivated in British Gardens,* Vol. 1. Published by the author and printed by James Moyes, Hatton Garden, London.

Hooker, W. 1989. *Hooker's Finest Fruits: A Selection of Paintings of Fruits by William Hooker 1779–1832.* Introduction by William T. Stearn, descriptions by Frederick A. Roach. Herbert Press in association with the Royal Horticultural Society, London.

Hopf, M. 1973. Apfel (*Malus communis* L.); Aprikose (*Prunus armeniaca* L.), in H. Beck, H. Jankuhn, K. Ranke, and R. Wenskus, editors, *Reallexikon der Germanischen Altertumskunde* [dictionary of German archaeology], Vol. 1.

Hopkirk, P. 1980. *Foreign Devils on the Silk Road: The Search for the Lost Treasures of Central Asia.* Murray, London. (1984 reprint, Oxford University Press)

Hosaka, K. 1995. Successive domestication and evolution of the Andean potatoes as revealed by chloroplast DNA restriction endonuclease analysis. *Theoretical and Applied Genetics* 90: 356–363.

Howard-Bury, C. 1990. *Mountains of Heaven: Travels in the Tien Shan Mountains, 1913.* M. Keaney, editor. Hodder and Stoughton, London.

Huckins, C. A. 1972. *A Revision of the Sections of the Genus Malus Miller.* Ph.D. thesis, Cornell University, Ithaca, New York.

Huckleberry, G., J. K. Stein, and P. Goldberg. 2003. Determining the provenience of Kennewick Man skeletal remains through sedimentological analyses. *Journal of Archaeological Science* 30: 651–665.

Hughes, W. 1672. *The American Physitian; or, a Treatise of the Roots, Plants, Trees, Shrubs, Fruit, Herbs, &c. Growing in the English Plantations in America.* London.

Ishikawa, S., S. Kato, S. Imakawa, T. Mikami, and Y. Shimamoto. 1992. Organelle DNA polymorphism in apple cultivars and rootstocks. *Theoretical and Applied Genetics* 83: 963–967.

Izhaki, I., and U. Safriel. 1989. Why are there so few exclusively frugivorous birds? Experiments on fruit digestibility. *Oikos* 54: 23–32.

Jacomet, S. 2005. Plant economy of the late 4th millennium BC cal in the northern Alpine foreland. *XVII International Botanical Congress—Abstracts:* 76. Vienna.

Janes, R. 1998. *Catalogue of Cultivars in the United Kingdom: National Fruit Collection.* Brogdale Horticultural Trust, printed at Wye College, London.

Janick, J., editor. 2003. Wild apple and fruit trees of Central Asia. *Horticultural Reviews* 29.

Janick, J., J. N. Cummins, S. K. Brown, and M. Hemmat. 1996. Apples, in J. Janick and J. N. Moore, editors, *Fruit Breeding,* Vol. 1, *Tree and Tropical Fruits.* Wiley, New York.

Janson, H. F. 1996. *Pomona's Harvest: An Illustrated Chronicle of Antiquarian Fruit Literature.* Timber Press, Portland, Oregon.

Janzen, D. H. 1982. Differential seed survival and passage rates in cows and horses, surrogate Pleistocene dispersal agents. *Oikos* 38: 150–156.

Jia Sixie. 1982. *Qi Min Yao Shu* [necessary skills for the masses, by a prefect of Gaoyang in Hebei, A.D. 534–550, edited and annotated by Miao Qiyu with Miao Guilong]. Agricultural Press, Beijing.

Jones, C. J., K. J. Edwards, S. Castaglione, W. O. Winfield, F. Sala, C. van de Wiel, G. Bredmijer, B. Vosman, M. Matthes, A. Daly, R. Brettschneider, P. Bettini, M. Buiatta, E. Maestri, A. Malcevschii, N. Marmiroli, R. Aert, G. Volckaert, J. Rueda, R. Linacero, A. Vazquez, and A. Karn. 1997. Reproducibility testing of RAPD, AFLP and SSR markers in plants by a network of European laboratories. *Molecular Breeding* 3: 381–390.

Jonston (sometimes spelled Jonstonus), J. 1662. *Dendrographias sive Historiae Naturalis de Arboribus et Fructibus tam Nostri Quam Peregrini Orbis Libri X* [writings concerning trees, 10 books of natural history concerning trees and fruits, both of our own and foreign regions]. Frankfurt.

Juniper, B. E. 1995. Waxes on plant surfaces and their interactions with insects. In R. J. Hamilton, editor, *Waxes: Chemistry, Molecular Biology and Functions.* Oily Press, Dundee.

Juniper, B. E., and S. B. Juniper, editors. 2003. *The Compleat Planter & Cyderist: Or, Choice Collections and Observations for the Propagating All Manner of Fruit-Trees, . . . by a Lover of Planting. . . . 1685.* Published by the editors, Oxford and Dursley, Gloucestershire.

Juniper, B. E., R. Watkins, and S. A. Harris. 1999. The origin of the apple. *Acta Horticulturae* 484: 27–33.

Kalkman, C. 2004. Rosaceae, pages 343–386 in K. Kubitzki, editor, *The Families and Genera of Vascular Plants,* Vol. 6, Flowering Plants. Dicotyledons. Celastrales, Oxalidales, Rosales, Cornales, Ericales. Springer, Berlin.

Karlgren, K. B. J. 1940. Grammatica Serica. Script and Phonetics in Chinese and Sino-Japanese. *Museum of Far Eastern Antiquities, Bulletin* 12: 1–471.

Kennedy, T. 1997. Old apples take root in national collection. *Technology Ireland* 29: 16–18.

Kitahara, K., S. Matsumoto, T. Yamamoto, J. Soejima, T. Kimura, H. Komatsu, and K. Abe. 2005. Parent identification of eight apple cultivars by S-RNase analysis and simple sequence repeat markers. *HortScience* 40: 314–317.

Knight, R. C., J. Amos, R. G. Hatton, and A. W. Witt. 1928. The vegetative propagation of fruit tree rootstocks. *Report of the East Malling Research Station* 14 and 15 for 1926–1927, Supplement 2: 11–30.

Knight, T. A. 1797. *A Treatise on the Culture of Apple & Pear, and on the Manufacture of Cider and Perry.* London. (other editions to 1818)

Knight, T. A. 1811. *Pomona Herefordensis: Containing Coloured Engravings of the Old Cider and Perry Fruits of Herefordshire. With Such New Fruits as Have Been Found to Possess Superior Excellence.* London.

Knoop, J. H. 1758. *Pomologia, Dat Is Beschryvingen en Afbeeldingen van der Beste Soorten van Appels en Peeren, Welke Neder- en Hoohg-Duitsland, Frankryk, Engelland en Elders Geagt Zyn, en Tot Dien Einde Gecultiveert Worden* [pomology, in which you will find descriptions and pictures of the varieties of apples and pears that are highly esteemed in Low- and High-Germany, France, England, and elsewhere, and are therefore for that reason cultivated]. Ferwerda, Leeuwarden.

Ko, K., J. L Norelli, J. P. Reynoird, S. K. Brown, and H. S. Aldwinckle. 2002. T4 lysozyme and attacin genes enhance resistance of transgenic 'Galaxy' apples against *Erwinia amylovora. Journal of the American Society for Horticultural Science* 127: 515–519.

Kobel, F. P., P. Steinegger, and J. Anliker. 1939. Weitere Untersuchungen über die Befruchtungsverhältnisse der Apfel- und Birnsorten [further researches on pollination behavior in varieties of apples and pears]. *Landwirtschaftliches Jahrbuch der Schweiz* 53: 160–191.

Korban, S. S., and R. M. Skirvin. 1984. Nomenclature of the cultivated apple. *Hort-Science* 19: 177–180.

Lack, H. W., with D. J. Mabberley. 1999. *The Flora Graeca Story: Sibthorp, Bauer and Hawkins in the Levant.* Oxford University Press.

Lamarck, continué par le citoyen J. M. Poiret. 1804. Pommier. *Malus,* Vol. 5, pages 559–563 in *Encyclopédie Methodique Botanique* [a methodical botanical encyclopedia]. Agasse, Paris.

Lamb, H. H. 1995. *Climate, History and the Modern World.* Routledge, London.

Lamboy, W. F., J. Yu, P. L. Forsline, and N. F. Weeden. 1996. Partitioning of allozyme diversity in wild populations of *Malus sieversii* L. and implications for germplasm collections. *Journal of the American Society for Horticultural Science* 121: 982–987.

Landry, B. S., R. Q. Li, W. Y. Cheung, and R. L. Granger. 1994. Phylogeny analysis of 25 apple rootstocks using RAPD markers and tactical gene tagging. *Theoretical and Applied Genetics* 89: 847–852.

Lane-Fox, R. 1973. *Alexander the Great.* Allen Lane, London.

Langford, T. 1681a. *Plain and Full Instructions to Raise All Sorts of Fruit-Trees that Prosper in England. . . .* London. (other editions to 1699)

Langford, T. 1681b. *The Practical Planter of Fruit Trees.* London.

Langley, B. 1729. *Pomona: Or, the Fruit-Garden Illustrated.* London.

Lanner, R. M. 1996. *Made for Each Other: A Symbiosis of Birds and Pines.* Oxford University Press.

La Quintinye (sometimes spelled La Quintinie), J. de. 1690. *Instruction pour les Jardins Fruitiers et Potagers, avec un Traité des Orangers, Suivy de Quelques Reflexions sur l'Agriculture* [instructions for the orchard and vegetable plot, with a treatise on oranges, followed by some reflections on agriculture]. Barbin, Paris.

Lauremberg (sometimes spelled Laurenberg), P. 1631. *Horticultura, Libris II. Comprehensa; . . .* [comprehensive horticulture], 2 vols. Stromer und Reichenbach, Nürnberg.

Laurence (sometimes spelled Lawrence), J. 1716. *The Clergy-Man's Recreation: Shewing the Pleasure and Profit of the Art of Gardening,* 4th edition. Lintot, London.

Lawson, W. 1618. *A New Orchard and Garden . . . with the Country Housewifes Garden . . .* London. (another edition 1623)

Leith-Ross, P. 1984. *The John Tradescants: Gardeners to the Rose and Lily Queen.* Owen, London.

Leroy, A. 1873. *Dictionnaire de Pomologie Contenant l'Histoire, la Description, la Figure des Fruits Anciens et des Fruits Modernes les Plus Généralement Connus et Cultivés* [a dictionary of pomology with the history, descriptions, and figures of the best known and most widely cultivated ancient and modern fruits], Vol. 3, Pommes. Angers, Paris.

Lespinasse, Y., and H. S. Aldwinckle. 2000. Breeding for resistance to fire blight, pages 253–273 in J. L. Vanneste, editor, *Fire Blight: The Disease and Its Causative Agent Erwinia amylovora*. CABI Publishing, New York.

Lev-Yadun, S., A. Gopher, and S. Abbo. 2000. The cradle of agriculture. *Science* 288: 1,602–1,603.

Lewin, R. 1998. Young Americans. *New Scientist* 2,156: 24–28.

Lewin, R. A. 1999. *Merde* [ordure]. Aurum Press, London.

Li Fan. 1984. *Zhongguo Zaipei Zhiwu Fazhan Shi* [a history of the development of cultivated plants in China]. Agricultural Press, Beijing.

Likhonos, A. 1974. A survey of the species in the genus *Malus* Mill. *Trudy po Prikladnoj Botanike, Genetike i Selektsii* 52: 16–34.

Lion, B. 1992. Vignes au royaume de Mari [vines of the kingdom of Mari]. *Mémoires de Nouvelles Assyriologies Brèves et Utilitaires* 1: 107–113.

Li Yunong. 1989. An investigation of the genetic center of *M. pumila* and *Malus* in the world. *Acta Horticulturae Sinica* 16: 101–108. (in Chinese with English abstract)

Li Yunong. 1996a. A critical review of the species and the taxonomy of *Malus* Mill. in the world. *Journal of Fruit Science* 13 (suppl.): 63–81.

Li Yunong. 1996b. A primarily modern systematics of the genus *Malus* Mill. in the world. *Journal of Fruit Science* 13: 83–91.

Li Yunong. 1999. An investigation and studies on the origin and evolution of *Malus domestica* Borkh. in the world. *Acta Horticulturae Sinica* 26: 203–220.

London, G., and H. Wise. 1699a. *The Compleat Gard'ner . . . by Monsieur de La Quintinye. Now Compendiously Abridg'd, and Made of More Use, with Very Considerable Improvements*. London. (abridged version of 1st edition of 1693; other editions to 1717)

London, G., and H. Wise. 1699b. *Fruit Walls Improved by Inclining Them to the Horizon*. London.

Longus. 1989. *Daphnis and Chloe*, pages 333–334 in B. P. Reardon, editor, *Collected Ancient Greek Novels*. University of California Press, Berkeley.

Loudon, J. C. 1844. *Arboretum et Fruticetum Britannicum; or, the Trees and Shrubs of Britain*, 8 vols. Longman, Brown, Green and Longman, London.

Lover of Planting, A. 1685. *The Compleat Planter & Cyderist: or, Choice Collections and Observations for the Propagating All Manner of Fruit-Trees, and the Most Approved Ways and Methods Yet Known for the Making and Ordering of Cyder, and other English-Wines*. London. (also see Juniper and Juniper 2003)

Luby, J., P. Forsline, H. Aldwinckle, V. Bus, and M. Geibel. 2001. Silk Road apples—Collection, evaluation, and utilization of *Malus sieversii* from Central Asia. *HortScience* 36: 225–231.

Lucie-Smith, E. 2001. *Flora: Gardens and Plants in Art and Literature*. Taschen, Köln.

Lucretius (Titus Lucretius Carus). 1907. Page 149 in *De Rerum Natura* [concerning natural things], translated by H. A. J. Munro. Bell, London, and Deighton Bell, Cambridge.

Mabberley, D. J. 1992. *Tropical Rain Forest Ecology,* 2nd edition. Blackie, Glasgow.

Mabberley, D. J. 2001. An Australian apple for Linnaeus. *The Gardens* (newsletter of the Friends of the Royal Botanic Gardens Sydney) 51: 11.

Mabberley, D. J. 2002. *Potentilla* and *Fragaria* reunited. *Telopea* 9: 793–801.

Mabberley, D. J. 2004. *Citrus* (Rutaceae): A review of recent advances in etymology, systematics and medical applications. *Blumea* 49: 481–498.

Mabberley, D. J., C. E. Jarvis, and B. E. Juniper. 2001. The name of the apple. *Telopea* 9: 421–430.

Maberly Family. 2005. (see *www.maberly.net*)

McGahan, J. 2001. Another apple for Snow White. *Pomona* 34: 10–11.

McLean, T. 1981. *Medieval English Gardens.* Collins, London.

McWhirter, A., and E. Clasen. 1996. *Foods That Harm and Foods That Heal.* Reader's Digest Association, London, New York, and Sydney.

Mallory, J. P., and V. H. Mair. 2000. *The Tarim Mummies.* Thames and Hudson, London.

Manganaris, A. G., F. H. Alston, N. F. Weeden, H. S. Aldwinckle, H. L. Gustafson, and S. K. Brown. 1994. Isozyme locus *Pgm*-1 is tightly linked to a gene (V_f) for scab resistance in apple. *Journal of the American Society for Horticultural Science* 119: 1,286–1,288.

Markham, G. 1625. *A Way to Get Wealth.* Lawson, London.

Markham, G. 1640. *The English Husbandman,* numerous separate parts, including Part 9, The Countryman's Recreation, or the Art of Planting, Grafting, and Gardening. London.

Marsh, R. W. 1983. *The National Fruit and Cider Institute: 1903–1983.* Annual Report of the Long Ashton Research Station.

Marshall, D. L., and D. M. Oliveras. 2001. Does differential seed siring success change over time or with pollination history in wild radish, *Raphanus sativus* (Brassicaceae)? *American Journal of Botany* 88: 2,232–2,242.

Martin, C. 2000. *A History of Canadian Gardening.* McArthur, Toronto.

Mascall, L. 1572. *A Booke of the Arte and Maner, Howe to Plante and Graffe All Sortes of Trees . . . by One of the Abbey of S. Vincent in Fraunce* [that is, Davy Brossard] . . . *Set Forth and Englished by Leonarde Mascal,* 2nd edition. London. (other editions 1569–1656)

Maslin, M. A., X. S. Li, M.-F. Loutre, and A. Berger. 1998. The contribution of orbital forcing to the progressive intensification of northern hemisphere glaciation. *Quaternary Science Reviews* 17: 411–426.

Matheus, P. E. 1995. Diet and co-ecology of Pleistocene short-faced bears and brown bears in eastern Beringia. *Journal of Quaternary Research* 44: 447–453.

Mattson, D. J., K. C. Kendall, and D. P. Reinhart. 2000. Whitebark pine, grizzly bears, and red squirrels, pages 121–136 in D. F. Tomback, S. F. Arno, and R. E. Keane, editors, *Whitebark Pine Communities: Ecology and Restoration.* Island Press, Washington and London.

Maund, B. 1845–1851. *The Fruitist.* Groombridge, London.

Meager, L. 1670. *The English Gardener: Or, a Sure Guide to Young Planters and Gardeners . . . of Planting All Sorts of Stocks, Fruit-Trees, and Shrubs.* London. (other editions to 1710)

Meiggs, R. 1982. *Trees and Timber in the Ancient Mediterranean World.* Clarendon Press, Oxford.

Meiggs, R., and D. A. Lewis. 1989. *A Selection of Greek Historical Inscriptions to the End of the Fifth Century B.C.* Clarendon Press, Oxford.

Merryweather, R. 1992. *The Bramley: A World Famous Cooking Apple.* Newark and Sherwood District Council, Nottingham.

Migot, A. 1957. *Tibetan Marches,* translated from the French by Peter Fleming. Penguin Books, Harmondsworth.

Miller, P. 1731. *The Gardeners Dictionary; Containing the Methods of Cultivating . . . .* London. (other editions to 1771)

Moerman, D. E. 1998. *Native American Ethnobotany.* Timber Press, Portland, Oregon.

Moore, G. A. 2001. Oranges and lemons: Clues to the taxonomy of *Citrus* from molecular markers. *Trends in Genetics* 17: 536–540.

Morgan, J. 1982. Vintage apples. *The Garden* (journal of the Royal Horticultural Society) 107: 308–313.

Morgan, J. 1993. Fruit history: The 'Decio' apple. *Fruit News* Autumn: 6–7.

Morgan, J., and A. Richards. 1993. *The Book of Apples.* Brogdale Horticultural Trust in association with Ebury Press, London.

Morgan, D. R., D. E. Soltis, and K. R. Robertson. 1994. Systematic and evolutionary implications of *rbc*L sequence variation in Rosaceae. *American Journal of Botany* 81: 890–903.

Morton Shand, P. 1949. Older kinds of apples. *Journal of the Royal Horticultural Society* 74: 60–67, 88–97.

Nazaroff, P. 1993. *Hunted Through Central Asia,* translated by Malcolm Burr. Oxford University Press.

Noiton, D. A. M., and P. A. Alspach. 1996. Founding clones, inbreeding, coancestry, and status number of modern apple cultivars. *Journal of the American Society for Horticultural Science* 121: 73–782.

Norelli, J. L., and H. S. Aldwinckle. 2000. Transgenic resistant varieties and root-stocks resistant to fire blight, pages 275–292 in J. L. Vanneste, editor, *Fire Blight: The Disease and Its Causative Agent Erwinia amylovora*. CABI Publishing, New York.

Nybom, H. 1990a. Genetic variation in ornamental apple trees and their seedlings (*Malus*, Rosaceae) revealed by DNA 'fingerprinting' with the M13 repeat probe. *Hereditas* 113: 17–28.

Nybom, H. 1990b. DNA fingerprints in sports of 'Red Delicious' apples. *HortScience* 25: 1,641–1,642.

Nybom, H., and B. A. Schaal. 1990. DNA "fingerprints" applied to paternity analysis in apples (*Malus* ×*domestica*). *Theoretical and Applied Genetics* 79: 763–768.

Nybom, H., S. H. Rogstad, and B. A. Schaal. 1990. Genetic variation detected by the use of the M13 'DNA fingerprint' probe in *Malus*, *Prunus* and *Rubus* (Rosaceae). *Theoretical and Applied Genetics* 79: 153–156.

Olsen, K. M., and B. A. Schaal. 1999. Evidence on the origin of cassava: Phylogeography of *Manihot esculenta*. *Proceedings of the National Academy of Sciences U. S.A.* 96: 5,586–5,591.

Oraguzie, N. C., H. C. M. Basset, M. Stefanati, R. D. Ball, V. G. M. Bus, and A. G. White. 2001. Genetic diversity and relationships in *Malus* sp. germplasm collections as determined by randomly amplified polymorphic DNA. *Journal of the American Society for Horticultural Science* 126: 318–328.

Oraguzie, N. C., T. Yamamoto, J. Soejima, T. Suzuki, and H. N. de Silva. 2005. DNA fingerprinting of apple (*Malus* spp.) rootstocks using simple sequence repeats. *Plant-Breeding* 124: 197–202.

O'Toole, C. 2000. *The Red Mason Bee: Taking the Sting Out of Bee-Keeping*. Osmia Publications, Banbury.

Page, D. 1955. *An Introduction to the Study of Ancient Lesbian Poetry: Sappho and Alcaeus*. Clarendon Press, Oxford.

Pakeman, R. J., G. Dignefe, and J. L. Small. 2002. Ecological correlates of endozoochory by herbivores. *Functional Ecology* 16: 296–304.

Palmer, J. W., J. P. Privé, and D. S. Tustin. 2003. Temperature, pages 217–236 in D. C. Ferree and I. J. Warrington, editors, *Apples: Botany, Production, and Uses*. CABI Publishing, Cambridge, Massachusetts.

Palmer, R. 1996. *Ripest Apples: An Anthology of Verse, Prose and Song*. Big Apple Association, Woodcroft, Putley, Ledbury, Herefordshire.

Palter, R. 2002. *The Duchess of Malfi's Apricots and Other Literary Fruits*. University of South Carolina Press, Columbia.

Pannell, C. M., and M. J. Kozioł. 1987. Arillate seeds and vertebrate dispersal in *Aglaia* (Meliaceae): A study of ecological and phytochemical diversity. *Philosophical Transactions of the Royal Society*, B, 316: 303–313.

Parkinson, J. 1629. *Paradisi in Sole Paradisus Terrestris* [of a paradise in the sun, a paradise of the earth]. London. (other editions to 1656)

Pelletier, A. 1975. Viticulture et oléiculture en pays allobroge dans l'antiquité: À propos du calendrier rustique de Saint-Romain-en-Gal. *Cahiers d'Histoire* 20: 21–26.

Pennington, R. T. 1996. Molecular and morphological data provide phylogenetic resolution at different hierarchical levels in *Andira*. *Systematic Biology* 45: 494–513.

Petzold. H. 1990. *Apfelsorten: Aquarelle und Zeichnungen* [varieties of apples: water-color paintings and descriptions], 2nd edition. Neumann, Leipzig-Radebeul.

Philips, J. 1708. *Cyder: A Poem in Two Books.* London. (other editions to 1791)

Phipps, J. B. 2003. *Hawthorns and Medlars.* Timber Press, Portland, Oregon.

Phipps, J. B., K. R. Robertson, P. G. Smith, and J. R. Rohrer. 1990. A checklist of the subfamily Maloideae (Rosaceae). *Canadian Journal of Botany* 68: 2,209–2,269.

Phipps, J. B., K. R. Robertson, J. R. Rohrer, and P. G. Smith. 1991. Origins and evolution of subfam. Maloideae (Rosaceae). *Systematic Botany* 16: 303–332.

Pliny (Gaius Plinius Secundus). 1967. *Historia Naturalis* [natural history], translated by H. Rackham. Heinemann, London.

Pollan, M. 2001. *The Botany of Desire: A Plant's-Eye View of the World.* Random House, New York.

Pollard, A., and F. W. Beech. 1957. *Cider-Making.* Hart Davis, London.

Ponomarenko, V. V. 1986. Obsor vidov roda *Malus* Mill. [a review of the species of the genus *Malus* Miller]. *Trudy po Prikladnoj Botanike, Genetike i Selektsii* 106: 16–27.

Ponomarenko, V. V. 1990. Wild apple in the central Kopet-Dag. *Trudy po Prikladnoj Botanike, Genetike i Selektsii* 134: 117–121.

Popov, M. G. 1929. [Wild growing fruit trees and shrubs of Asia Media]. *Trudy po Prikladnoj Botanike, Genetike i Selektsii* 22: 241–483.

Popova, G., and M. G. Popov. 1925. [The wild apple tree in the valley of Tchimgan (western Tianschan)]. *Bjulleten' Sredne-Aziatskogo Gosudarstvennogo Universiteta, Bulletin de l'Université de l'Asie Centrale* 11: 99–103.

Postgate, J. N. 1987. Notes on fruit in the cuneiform sources. *Bulletin for Sumerian Agriculture* 3: 115–120, 128–132.

Potts, D. T. 2004. Camel hybridization and the role of *Camelus bactrianus* in the ancient Near East. *Journal of the Economic and Social History of the Orient* 47: 143–165.

Power, B., and E. Cocking. 1991. How we saved the Bramley. *The Garden* (journal of the Royal Horticultural Society) 116: 90–94.

Raphael, S. 1990. *An Oak Spring Pomona: A Selection of the Rare Books on Fruit in the Oak Spring Garden Library Upperville, Virginia.* Yale University Press, New Haven.

Rea, J. 1665. *Flora: Seu, de Florum Cultura. . . . (Flora.–Ceres.–Pomona),* . . . [plants: or the culture of plants . . . flowers–agriculture–fruits]. London. (other editions to 1702)

Regan, B. C., C. Julliot, B. Simmen, F. Viénot, P. Charles-Dominique, and J. D. Mollon. 2001. Fruits, foliage and the evolution of primate colour vision. *Philosophical Transactions of the Royal Society,* B, 356: 229–283.

Rehder, A. 1926. *Manual of Cultivated Trees and Shrubs.* Macmillan, New York.

Remmy, K. von, and F. Gruber. 1993. Untersuchungen zur Verbreitung und Morphologie des Wild-Apfels [researches on the distribution and morphology of the wild apple] (*Malus sylvestris*) (L.) Mill.). *Mitteilungen der Deutschen Dendrologischen Gesellschaft* 81: 71–94.

Renard, C. M., A. Baron, S. Guyot, and J.-F. Drilleau. 2001. Interactions between cell walls and native polyphenols: Quantification and some consequences. *International Journal of Biological Macromolecules* 29: 115–125.

Renfrew, C. 1989. *Archaeology and Language: The Puzzle of Indo-European Origins.* Penguin, London.

Renfrew, J. M. 1987. The archaeological evidence for the domestication of plants: Methods and problems, pages 149–172 in P. J. Ucko and G. W. Dimbleby, editors, *The Domestication and Exploitation of Plants and Animals.* Duckworth, London.

Richards, A. J. 1986. *Plant Breeding Systems.* Allen and Unwin, London.

Richthofen, Baron F. von. 1883. *China: Ergebnisse Eigener Reisen und Darauf Gegrundeter Studien, 1877–1882* [China: the results of my various travels and the studies based thereon], Vol. 4 (of 5). Berlin.

Rivers, T. 1870. *The Miniature Fruit Garden.* Longman, Brown, Green and Longman, London.

Roach, F. A. 1985. *Cultivated Fruits of Britain.* Blackwell, Oxford.

Robb-Smith, A. H. T. 1956. Blenheim Orange variants, false Blenheims and Blenheim seedlings, pages 1–23 in P. M Synge and L. Roper, editors, *The Fruit Year Book* 9. Royal Horticultural Society, London.

Roberts, R. H. 1949. Theoretical aspects of graftage. *Botanical Review* 15: 423–463.

Robinson, J. P., and S. A. Harris. 2000. Amplified fragment length polymorphisms and microsatellites: A phylogenetic perspective, pages 95–121 in E. M. Gillet, editor, *Which DNA Marker for Which Purpose?* Institut für Forstgenetik und Forstpflanzenzüchtung, Göttingen.

Robinson, J. P., S. A. Harris, and B. E. Juniper. 2001. Taxonomy of the genus *Malus* Mill. (Rosaceae) with emphasis on the cultivated apple, *Malus domestica* Borkh. *Plant Systematics and Evolution* 226: 35–58.

Rohrer, J. R., K. R. Robertson, and J. B. Phipps. 1994. Floral morphology of Maloideae (Rosaceae) and its systematic relevance. *American Journal of Botany* 81: 574–581.

Rom, R. C., and R. F. Carlson, editors. 1987. *Rootstocks for Fruit Crops*. John Wiley and Sons, New York.

Room, A. 1998. *Brewer's Dictionary of Phrase and Fable*, 15th edition. Cassell, London.

Ruel, J. 1536. *De Natura Stirpium Libri Tres* [concerning natural features of plants in three books]. Paris.

Sanders, R. 1988. *The English Apple*. Royal Horticultural Society in association with Phaidon Press, London.

Sargent, C. S. 1922. *Manual of the Trees of North America (Exclusive of Mexico)*, 2nd edition. (1961 reprint, Dover, New York)

Savolainen, V., R. Corbaz, C. Moncousin, R. Spichiger, and J.-F. Manenj. 1995. Chloroplast DNA variation and parentage analysis in 55 apples. *Theoretical and Applied Genetics* 90: 1,138–1,141.

Sax, K. 1931. The origin and relationships of the Pomoideae. *Journal of the Arnold Arboretum* 12: 3–22.

Schiemann, E. 1932. Entstehung der Kulturpflanzen [origin of cultivated plants], in E. Baur and M. Hartmann, editors, *Handbuch der Vererbungswissenschaft* [handbook of the science of inheritance], Vol. 3. Borntraeger, Berlin.

Schmidt, W. C., and F. K. Holtmeier, editors. 1994. *Proceedings—International Workshop on Subalpine Stone Pines and Their Environment: The Status of Our Knowledge*. U.S. Forest Service General Technical Report INT-GTR-309. Ogden, Utah.

Schneider, C., and C. Moritz. 1999. Rainforest refugia and evolution in Australia's wet tropics. *Proceedings of the Royal Society of London*, B, 266: 191–196.

Schneider, C. J., T. B. Smith, B. Larison, and C. Moritz. 1999. A test of alternative models of diversification in tropical rainforests: Ecological gradients vs. rain forest refugia. *Proceedings of the National Academy of Sciences U.S.A.* 96: 13,869–13,873.

Schuyler, E. 1876. *Turkistan: Notes of a Journey in Russian Turkistan, Khokand, Bukhara, and Kuldja*, Vol. 1. Sampson Low, Marston, Searle & Rivington, London.

Schweingruber, F. H. 1979. Wildapfel und Prähistorische Apfel [the wild apple and the prehistoric apple]. *Archaeo-Physika* 8: 283–294.

Scott, J. 1873. *The Orchardist or Catalogue of Fruits Cultivated at Merriott, Somerset*. Pollett, London.

Sherratt, A. 1984. History and the horse. *Ashmolean* (5): 4–7.

Sherratt, A. 2004. The horse and the wheel: The dialectics of change in the circum-Pontic region and adjacent areas, 4,500–1,500 B.C., pages 233–252 in M. Levine, C. Renfrew, and K. Boyle, editors, *Prehistoric Steppe Adaptation and the Horse: Ancient Interactions: East and West in Eurasia*. McDonald Institute Monographs, Cambridge.

Shouse, B. 2001. Spreading the word, scattering the seeds. *Science* 294: 988–989.

Simmonds, A. 1946. *A Horticultural Who Was Who.* Royal Horticultural Society, London.

Simon Louis-Frères. 1896. *Catalogue.* Metz.

Simoons, F. 1991. *Food in China.* CRC Press, Boca Raton, Florida.

Smith, M. W. G. 1971. *National Apple Register of the United Kingdom.* Ministry of Agriculture, Fisheries and Food, Castle Point Press, Dunbeatty, Scotland.

Society for New York City History. 1995. Why is New York City called "The Big Apple"? (see *http://salwen.com/apple.html*)

Soest, L. J. M. van, K. I. Baimatov, V. F. Chapurin, and A. P. Pimakhov. 1998. Multicrop collecting mission to Uzbekistan. *Plant Genetic Resources Newsletter* 116: 32–35. (short communication)

Spiers, V. 1996. *Burcombes, Queenies and Colloggetts: The Makings of a Cornish Orchard,* illustrated by Mary Martin. West Brendon.

Stafford, H. 1755. *A Treatise of Cyder Making.* London.

Stebbins, G. L. 1950. *Variation and Evolution in Plants.* Columbia University Press, New York.

Stein, M. A. 1903. *Sand-Buried Ruins of Khotan.* Unwin, London.

Stein, M. A. 1907. *Ancient Khotan,* 2 vols. Clarendon Press, Oxford.

Stein, M. A. 1921. *Serindia,* 4 vols. Clarendon Press, Oxford.

Stevens, P. F. 1991. George Bentham and the Kew Rule, pages 157–168 in D. L. Hawksworth, editor, Improving the stability of names: Needs and options. *Regnum Vegetabile* 123.

Switzer, S. 1724. *The Practical Fruit-Gardener.* London. (other editions to 1763)

Szewc-McFadden, A. K., A. K. S. Bleik, C. G. Alpha, W. F. Lamboy, and J. R. McFerson. 1995. Identification of simple sequence repeats in *Malus* (apple). *HortScience* 30: 855. (abstract)

Szewc-McFadden, A. K., W. F. Lamboy, S. C. Hokanson, and J. R. McFerson. 1996. Utilization of identified simple sequence repeats (SSRs) in *Malus* ×*domestica* (apple) for germplasm characterization. *HortScience* 31: 619. (abstract)

Tallents, S. 1956. The Sir Isaac Newton apple. *Fruit Year Book* 9: 35–40. Royal Horticultural Society, London.

Taylor, D. 1999. Ornithological Society of the Middle East trip report: Kazakhstan, May 27–June 5, 1999. (see *www.osme.org*)

Taylor, H. V. 1948. *The Apples of England.* Crosby, Lockwood, London.

Theophrastus. 1916. *Enquiry into Plants,* 2 vols. Loeb Classical Library, London and Cambridge, Massachusetts.

Thesiger, W. 1979. *Desert, Marsh and Mountain: The World of a Nomad.* HarperCollins, London.

Tobutt, K. R., R. Bošković, and P. Roche. 2000. Incompatibility and resistance to woolly apple aphid in apple. *Plant Breeding* 119: 65–69.

Traveset, A. 1998. Effect of seed passage through vertebrate frugivores' guts on germination: A review. *Perspectives in Plant Ecology, Evolution and Systematics* 1: 151–190.

Traveset, A., and M. F. Wilson. 1997. Effect of birds and bears on seed germination of fleshy-fruited plants in temperate rainforests of southeast Alaska. *Oikos* 80: 89–95.

Tukey, H. B. 1964. *Dwarfed Fruit Trees for Orchard, Garden and Home.* Collier-Macmillan, London and New York.

Tusser, T. 1557. *A Hundreth Good Pointes of Husbandrie.* London.

Tusser, T. 1573. *Five Hundred Points of Good Husbandrie United to as Many of Good Huswiferie.* London. (also later editions)

Twiss, S. 1999. *Apples: A Social History.* National Trust, London.

Tydeman, H. M. 1937. The wild fruit trees of the Caucasus and Turkestan: Their potentialities as rootstocks for apples and pears. 1. A first report on some wild quinces from the Caucasus. *East Malling Research Station Annual Report* 103–116.

Ucko, P. J., and G. W. Dimbleby, editors. 1969. *The Domestication and Exploitation of Plants and Animals.* Duckworth, London.

Vavilov, N. I. 1926. Studies on the origin of cultivated plants. *Trudy po Prikladnoj Botanike i Selektsii* 16: 139–245.

Vavilov, N. I. 1930. Wild progenitors of the fruit trees of Turkistan and the Caucasus and the problem of the origin of fruit trees, pages 271–286 in *Proceedings, International Horticultural Congress.* Royal Horticultural Society, London.

Vavilov, N. I. 1951. *The Origin, Variation, Immunity and Breeding of Cultivated Plants,* translated by K. Starr Chester. Chronica Botanica, Waltham, Massachusetts.

Vavilov, N. I. 1992. *Origin and Geography of Cultivated Plants,* translated by D. Löve. Cambridge University Press.

Venette, N. 1685. *The Art of Pruning Fruit-Trees . . . with an Explanation of Some Words Which Gardiners Make Use of in Speaking of Trees. And a Tract of the Use of the Fruits of Trees, for Preserving Us in Health, or for Curing Us When We Are Sick. Translated from the French Original, Set Forth in the Last Year by a Physician of Rochelle.* London.

Vila, C., J. A. Leonard, A. Gotherstrom, S. Marklund, K. Sandberg, K. Liden, R. K. Wayne, and H. Ellegren. 2001. Widespread origin of domestic horse lineages. *Science* 291: 474–477.

Villaret–von Rochow, M. 1969. Fruit size variability of Swiss prehistoric *Malus sylvestris,* pages 201–206 in P. J. Ucko and G. W. Dimbleby, editors, *The Domestication and Exploitation of Plants and Animals.* Duckworth, London.

Walker, A. 1998. *Aurel Stein: Pioneer of the Silk Road.* Murray, London.

Walters, S. M. 1961. The shaping of angiosperm taxonomy. *New Phytologist* 60: 74–84.

Ward, R. A. 1988. *Harvest of Apples.* Penguin, London.

Ward, R. 1992. Lord of the cider apples. *The Garden* (journal of the Royal Horticultural Society) 117: 512–513.

Wasserman, H., M. Pastoureau, M. Preaud, T. Ky, F. Drouard, R. Buren, and L. Lachenal. 1990. *La Pomme: Histoire Symbolique et Cuisine* [the apple: its symbolic history and cuisine]. Sang de la Terre, Paris.

Watkins, R. 1995. Apple and pear, pages 418–422 in J. Smartt and N. W. Simmonds, editors, *Evolution of Crop Plants.* Longman, London.

Way, R. D. 1976. The largest apple variety collection in the United States. *New York's Food Life Sciences* 9: 11–13.

Way, R. D., H. S. Aldwinckle, R. C. Lamb, A. Rejman, T. Sansavini, T. Shen, R. Watkins, M. N. Westwood, and Y. Yoshida. 1990. Apples (*Malus*), pages 1–62 in J. N. Moore and R. Ballington, editors, *Genetic Resources of Temperate Fruits and Nuts.* International Society of Horticultural Science, Leuven, Belgium.

Webster, A. D., and S. J. Wertheim. 2003. Apple rootstocks, pages 91–124 in D. C. Ferree and I. J. Warrington, editors, *Apples: Botany, Production, and Uses.* CABI Publishing, Cambridge, Massachusetts.

Welch, C. A., J. Keay, K. C. Kendall, and C. T. Robbins. 1997. Constraints on frugivory in bears. *Journal of Ecology* 78: 1,105–1,119.

Wen, J. 1999. Evolution of eastern Asian and eastern North American disjunct distributions in flowering plants. *Annual Review of Ecology and Systematics* 30: 421–455.

Wertheim, S. J., and A. D. Webster. 2003. Propagation and nursery tree quality, pages 125–151 in D. C. Ferree and I. J. Warrington, editors, *Apples: Botany, Production, and Uses.* CABI Publishing, Cambridge, Massachusetts.

Westwood, M. N. 1995. *Temperate Zone Pomology: Physiology and Culture,* 3rd edition. Timber Press, Portland, Oregon.

Whitaker, Reverend T. 1805. *History and Antiquities of the Deanery of Craven, in the County of York.* London.

Williams, R. R. 1987. *Cider and Juice Apples: Growing and Processing.* University of Bristol Press.

Williams, R. R., and R. D. Child. 1965. The identification of cider apples, pages 71–89 in *Long Ashton Research Station Annual Report 1965.*

Willis, K. J. 1996. Where did all the flowers go? The fate of temperate European flora during glacial periods. *Endeavour* 20: 110–114.

Wilson, E. H. 1913. *A Naturalist in Western China.* (1986 reprint, Cadogan Books, London)

Wiltshire, P. E. J. 1995. The effect of food processing on the palatability of wild fruits with high tannin content, pages 385–397 in H. Kroll and R. Pasternak, editors, *Res Archaeobotanicae International Workgroup for Paleoethnobotany 1992.* Oetker-Voges, Kiel.

Winterbottom, M., and Thomson, R. M. 2005. *Gesta Pontificum Anglorum* [the deeds of the bishops of England] *of William of Malmesbury c. 1125,* Vol. 1. Clarendon Press, Oxford.

Witt, R. de. 2000. Pomegranate, not apple, likely fruit in Garden of Eden. *World of Wood* August: 6–7.

Worlidge, J. 1669. *Systema Agriculturae, the Mystery of Husbandry Discovered.* London.

Worlidge, J. 1676. *Vinetum Britannicum: Or, a Treatise of Cider and Other Wines and Drinks Extracted from Fruits . . .* London. (other editions to 1691)

Wünsch, A., and J. I. Hormaza. 2002. Molecular characterization of sweet cherry (*Prunus avium* L.) genotypes using peach [*Prunus persica* (L.) Batsch] SSR sequences. *Heredity* 89: 55–63.

Yao, J.-L., Y.-H. Dong, and B. A. M. Morris. 2001. Parthenocarpic apple fruit production conferred by transposon insertion mutations in a MADS-box transcription factor. *Proceedings of the National Academy of Sciences U.S.A.* 98: 1,306–1,311.

Zemanek, A., and J. de Koning. 1998. Plant Illustrations in the *Libri Picturati* (A18–30) and new currents in Renaissance botany, pages 161–193 in Z. Mirek and A. Zemanek, editors, *Studies in Renaissance Botany.* Guidebook Series 20. Polish Academy of Sciences, Krakow.

Zhang, W., J. Zhang, and X. Hu. 1993. Distribution and diversity of *Malus* germplasm resources in Yunnan, China. *HortScience* 28: 978–980.

Zhou, Z.-Q. 1999. Apple genetic resources in China: The wild species and their distributions, informative characteristics and utilization. *Genetic Resources and Crop Evolution* 46: 599–609.

Zhou, Z.-Q., and Y.-N. Li. 2000. The RAPD evidence for the phylogenetic relationship of the closely related species of the cultivated apple. *Genetic Resources and Crop Evolution* 47: 353–357.

Zohary, D., and M. Hopf. 2000. *Domestication of Plants in the Old World.* Oxford University Press.

Zohary, D., and P. Spiegel-Roy. 1975. Beginnings of fruit growing in the Old World. *Science* 187: 319–327.

Index